What to Grow for a CLIMATE RESILIENT GARDEN

Alocasia micholitziana

RHS

What to Grow for a CLIMATE RESILIENT GARDEN

Dr Amanda Rasmussen

greenfinch

Contents

Introduction

The sound of friendly chatter mingles with birdsong and the buzz of bees as the sun beats down on the umbrella of tree branches shading the outdoor seating area. The conversation is (of course) about the weather: the surprisingly warm recent days, and the floods of just a few months ago.

The weather is a small-talk opener in many cultures, from pondering whether rain will fall any time soon, as is common in my Australian homeland, to wondering if it will ever stop, as in my current home of the UK. But while we all love to talk about the weather, such conversations are becoming a more serious concern in line with an increasing battering of storms, unusually hot weather or seemingly endless droughts.

It's important to remember that climate change is an increase in temperature measured over decades, rather than the differences that occur from year to year. Every decade since the 1980s has been warmer than the previous one. This is caused by greenhouse gases, such as carbon dioxide, trapping heat in the Earth's atmosphere. When the sun's rays approach the Earth, some bounce off layers in the atmosphere or the surface of our planet and are reflected back into space. Greenhouse gases prevent some of that reflected heat from leaving the atmosphere, trapping it in the way a greenhouse traps the warmth of the sun. This extra heat warms up the atmosphere and oceans, resulting in changes to the Earth's weather patterns.

For most places, this means that extreme weather events will become more frequent. Regions in the centre of large continents, far from the ocean, may experience more extreme droughts. As the air warms, more water can evaporate from the soil. Drier soil heats up more than wet soil, which further increases the air temperature above it, and this in turn means the air can hold more moisture, evaporating any that is left in the soil. Rain is unlikely to occur under these conditions until the temperature falls or a moisture-laden air mass blows in.

In areas near the ocean, by contrast, more water evaporates from the sea as the temperature rises. When this moisture-filled air moves over land and meets cooler air (such as that in mountains or from an air mass originating close to the poles), all that extra moisture drops out of the air as an intense downpour or storm. The warm ocean air also feeds the energy that builds in tropical storms, including hurricanes, cyclones and typhoons, increasing the force with which they slam the coastlines.

Local climates are also influenced by a range of other factors, including landcover type – forest, grassland, bare earth or cities, for example – and by local terrain, such as mountains, river valleys and plains. The exact changes that each location will experience will therefore vary according to the way local and global weather patterns interact in that particular place. But whatever the historical 'normal' climate of any location, it is very likely that more and more extreme weather events will confront us in the future.

While many of us are initially most concerned about the safety of our loved ones and property in extreme weather, our gardens and green spaces can either suffer from or help to mitigate the effects of climate change. Which of these occurs is down to our choice of landscaping and plants. Let me explain.

Sedges

What climate change means for gardening, and what gardening can do to help

Storms that bring pulses of high rainfall in urban areas can cause flash flooding when the gutters and drains are hit all at once with high volumes of water running off roofs and paved areas. Slowing the speed at which water enters the drains can avoid that sudden rush and prevent flash flooding – and this is where the garden comes in. Diverting even some of the rain from downpipes into water storage can help to reduce the volume of water in the drains. This will be even more effective if the water storage overflows into 'rain gardens' with both low and high areas. Water is diverted through the low areas and plants that prefer more moisture can grow there happily, while those that prefer not to have their feet wet can be grown in the higher spots. Rain gardens are also beneficial during drought, since the water is concentrated in the low areas, which remain moist for longer.

Green roofs are another great way to reduce the amount of water that reaches the drains, while also providing insulation for the building against extremes of temperature. In apartment blocks, flat roofs are increasingly being converted into green spaces that capture rainfall while providing recreation areas for the residents. If your apartment or office block has a flat roof, consider asking the building manager if you can start a rooftop garden. Just be sure to consider the weight of both the garden and the waterproofing or you might end up with leaks rather than leeks.

The appropriate choice of plants can also play a role when rainfall is high. Plants that can cope with wet feet will continue growing, whereas others will stop growing and rot away. Since healthy plants take up water from the soil and transpire (lose water into the atmosphere) as part of their growth, they reduce the soil water. Planting large-leaved plants that are happy in boggy conditions will minimize the damage that is caused during a wet season.

Gardens are not only important for averting floods; they can also reduce the effect of hot, dry weather. Perhaps the first benefit that comes to mind is shading. The less direct sun that hits a building, the lower the temperature inside it. Plants also cool the environment through the movement of water. As water evaporates from the leaves, it cools the plant, further buffering the area against heat. A sun-baked area between buildings, for example, can be turned from an oven into a pleasant oasis through the use of appropriate trees and shrubs.

Cities are often several degrees warmer than the surrounding countryside, forming so-called heat islands. This is because concrete buildings absorb heat from the sun and reflect it into the spaces that surround them. Similarly, climate change is increasing the average global temperature and causing more frequent periods of extreme conditions. Some native plant species will not be able to survive at these warmer temperatures or withstand the extreme weather events. If, for our gardens and city spaces, we choose plants that will thrive there, they will reduce the ambient temperature and buffer against weather extremes. Native plants will then be able to grow happily among these protective shrubs and trees.

Hero plants

Each plant species has its own superpowers to help it adapt to its environment, and those that are native to particularly challenging environments have some pretty impressive attributes. Chapter 1 explains how each type of garden superhero differs from the others in its growth and survival strategies. Chapters 2–9 highlight the superpowers of specific plants suited to particular environments, while Chapter 10 explores attractive plants that are adapted to any one of these environments, but that are also edible – the edimentals.

No matter what your climate, increasing plant diversity benefits the diversity of birds and insects and buffers against the impact of climate change. That diversity can consist of both native and sensibly chosen non-native plants, mixing evergreen and deciduous grasses, shrubs and trees. Climate change is increasing the risk that some plants become invasive and that plant pests or diseases spread to previously uninfected locations. This means that to be a responsible, environment-friendly gardener requires thought and consideration. But fear not; there are plenty of people to advise you, including the staff at your local nursery and botanic garden.

Limonium bourgeaui

Changing the way we think about plants

The word 'weed' is falling out of favour as gardeners become more open-minded about the plants they choose. It is used less for a particular, unchanging set of 'undesirable' plants than for species that make a nuisance of themselves in a particular place. The most common definition of 'weed' today is 'a plant out of place'; in a formal garden that might include native grasses or the seedlings of trees that will become far too large for the enclosure, whereas in nature reserves it refers to non-native plants, and sometimes those that are encroaching as a result of changes in the management of the reserve. To be defined as such, a weed does not have to be invasive, but plants that become invasive become the worst weeds for gardeners and native habitats alike.

Invasive plants are those that spread quickly and outcompete others. Such plants make efficient use of the various methods of spreading seed: through birds eating the seeds (as with *Lantana*), then flying away and pooping them out accompanied by a nice deposit of fertilizer ready for germination; using hooks that attach to the fur of passing animals, allowing the seeds to hitchhike; or by producing feather-light seed that floats away on the wind.

Plants can also spread by suckers and runners. Below ground, stems can grow sideways, protected from heat and cold, then send up new shoots next to the parent plant or quite a distance away. Bamboo is a good example of this method. Clumping bamboo varieties tend to shoot close to the parent plant, while spreading bamboos can send up shoots as far away as 9m (30ft).

Viola odorata

Japanese knotweed (*Reynoutria japonica*), a major invasive weed in areas of Europe and the USA, behaves in the same way.

Many aquatic species produce new plants from stems that break off and wash downriver. This means that anything planted in a water garden must be treated with great caution. If there is any chance that the garden will overflow into local creeks, canals or rivers, stick to plants that are native to the region.

It is easy to list the features that may result in a plant being considered a weed, but almost every species of plant will have one or more of those attributes, so context is everything. Plants that cannot tolerate frost, for example, are unlikely to become invasive in cold climates where the winters still freeze reliably. Likewise, a plant in a balcony water garden or wildlife pond surrounded by a dry landscape is very unlikely to escape and cause a problem. But if the local climate warms, or floods become more frequent, these plants may become invasive.

Being a responsible gardener

Remembering that diversity is the goal for habitat creation in climate resilient gardens, it's important to think about the plants we choose. Let's explore a worst-case scenario: think about the worst weed in your local nature reserve. Chances are it's a garden escapee – something that was brought in as an ornamental and is now running rampant in the wild. For me, the one that comes to mind is *Lantana camara*. In eastern Australia this prickly-stemmed, fuzzy-leaved nightmare, spread by birds, is smothering the edges of forests and spilling into grassland. Pulling it out and crushing the leaves releases a scent that is burned into my olfactory memory. In the UK, however, plant catalogues advertise the latest varieties of this member of the verbena family (Verbenaceae). How can this be? The reality is that this native of the American tropics is a benign (and arguably beautiful) cottage-garden plant in the UK, where the cooler climate means it is not at risk of escaping into the wild – for now.

This example is typical. It is impossible to suggest a single list of plants suitable for each environment in every location in the world – and how boring gardens would be if everyone planted the same! In this book, I advise caution where a plant is already considered invasive, and suggest local alternatives. My hope is that this book will provide a larger list of possible plants. To help, while many sections concentrate on a single species, similar species are also mentioned. I hope, too, that understanding the traits that make those plants well adapted to a particular environment will show you what to look for in other plants.

While all this sounds daunting, there are ways of minimizing the risk from invasive plants. Locally native species are often a safe choice and will support diverse wildlife. However, with climate change some of these will no longer be suited to their original locations. Other plants can be grown in pots first to see how they fare, or barriers can be dug into the garden to slow the spread of runners and suckers. Visit local public and botanic gardens to see what grows well in your area and observe what precautions are taken to minimize the spread of plants. Experts there or at a local nursery can also help with non-invasive choices. If you are still in doubt, don't take the risk; simply choose a different plant.

There are other aspects of being a responsible gardener. If you want to attract wildlife, it's important to minimize your use of pesticides. Providing safe access to water (explored in Chapter 2) benefits all wildlife, and keeping bird feeders clean avoids them spreading disease, such as bird flu, and parasites among birds.

It's also important to obtain plants responsibly. Never bring home seeds or plants from abroad, or buy plants online that come from overseas, since they could be carrying the microscopic spores of a disease or eggs of a pest that are not yet a problem in your home area, or they might have been dug up unscrupulously from the wild. Instead, purchase non-native plants from a reputable nursery, which will have followed the appropriate hygiene rules when importing the plants. Never take plants from the wild, as this can disrupt the local ecosystems. Instead, find a native nursery that will be able to supply the seeds or plants ethically.

When it comes to potting mix, go peat-free. Peat is made largely of plant material that has broken down in waterlogged and low-oxygen conditions, producing this fine, soft soil that holds moisture really well. However, it takes thousands of years to form, and fragile heath ecosystems are dependent on it. Instead, buy peat-free potting mixes and support nurseries that avoid peat.

The use of plastic is a big part of gardening today, so minimize it as much as you can. Where possible, use biodegradable pots. It may sometimes be necessary to use plastic – for seed trays, for example – but tough ones will last for many years, while cheap, thin ones will disintegrate quickly into microplastics that will pollute your garden for generations to come. Even better, choose tough products made from recycled plastic.

One much talked-about way of mitigating climate change is to plant a tree. But the most important mantra to remember is: right tree, right place. Many people get nervous about planting a tree in a small garden or close to the house. A giant oak next to the back door might not be the best choice, of course, but there are plenty of small trees that will not encroach on the foundations or underground utilities. Talk to the experts at your local tree nursery or botanic garden. It's also worth remembering that grassland is another hugely productive wildlife haven. Many butterflies rely on particular grasses as food for their caterpillars – so if wildlife is your interest, think about the creatures you want to attract and focus first on identifying the plants that will help.

We now understand that garden climates are changing, and have an idea of the responsibilities we need to think about. In the pages that follow you will find lots of exciting science, but why should we believe it? Scientists study plants and write up their findings in articles. These articles are then read and criticized by other scientists, and it is only when they pass this 'peer review' that these exciting findings are published. I have used these scientific papers to research the pages that follow, and a full list can be found on pages 216–19. So now let's learn more about what plants we should consider for the climate resilient garden.

Dicksonia antarctica

Cattleya

Chapter 1

Climate Change Science and Superhero Plants

Introduction
Climate Change Science and Superhero Plants

Why do 'normal' plants wilt in the summer sun? Why is a particular plant recommended for drought-resilient gardens? Why is another able to grow underwater?

The answers to these questions emerge when we begin to explore the amazing ways in which plants have adapted to their native environments. We can think of these as their superpowers. Every habitat poses challenges to the plants that live in it, so rather than there existing a 'normal' plant, every species has evolved strategies that help it to live in a particular habitat. Some environments pose more serious challenges than others. Chapters 2–9 each focus on a different one of these environments, exploring the adaptations that garden plants native to each habitat have evolved to help them overcome the challenges. But before we launch into those species, let's explore what makes plants so fascinating and diverse.

We know that all plants have the same structure of roots, stems and leaves, right?

WRONG!

There are many plants that do not have leaves, or don't produce stems, and even some that don't produce roots. We'll explore each of these in the following pages, but first let's introduce plants properly.

Ophrys apifera

Plant names

Scientific names are sometimes long and hard to pronounce, but they are the best way to know which plant is which. This is because common names vary from location to location, and the same name is often used for several species that might not be related – and that's just within a single language! Scientific names (also called botanical names) avoid this confusion because they do not change from place to place or between languages.

Each scientific name is presented in italics and made up of two parts. The first part is the genus (plural: genera), the plant's surname, if you like. The second is the species. For example, *Thymus serpyllum* is a plant in the genus *Thymus*, and is the species *serpyllum*. Some species are divided further into subspecies, such as *Thymus serpyllum* subsp. *tanaensis*. The genus name is conventionally shortened – to *T. serpyllum*, in this example – after it has been mentioned once, to save repetition.

Each genus belongs to a family of many genera. *Thymus* is in the family Lamiaceae. Genera within a family are more closely related to one another than to plants in a different family.

However, this doesn't necessarily mean that all plants within the same genus or family look the same. While there will be some anatomical details (such as the shape of particular details in the flower or seed) that are common to all members of a family or genus, many of the features we observe will depend on the habitat in which the plant evolved. Just as a whale (a mammal) looks similar to a shark (a fish), unrelated plants in similar habitats will develop similar traits (features) to help them thrive in that place.

Arachis hypogaea

Plant niches

The habitat in which a plant (or any organism) can live is called, more specifically, its niche. This is made up of the climate in the spot where the plant grows (also called the microclimate), the structure of that spot (perhaps on a solid branch in the tree canopy, or in sandy soil, or a boggy marsh), and the available resources (light, nutrients and water). The availability of resources can be influenced by other things, such as whether the soil is acid (as with heathland) or alkaline (chalky soils), whether there is salt in the soil, and what competing plants are growing nearby.

In a shaded spot under trees, it might seem as though we can grow only shade-loving plants, such as *Hosta*, but some plants have evolved different growth forms to adjust to the niche they experience. Let's think about *Clematis*, a popular garden climber. *Clematis* prefers to keep its roots cool in the shade, but will happily climb into the canopy to get more light for its leaves and flowers. Its niche, then, is defined as shaded or cool feet and higher light on the foliage.

When we choose plants for our gardens, it's helpful to think about each growth form and the niche to which it has adapted. The box opposite describes some of the different growth forms and gives one or more garden examples of each. We'll explore some of these as we uncover the superpowers associated with below-and above-ground plant parts.

Aeonium arboreum

Plant growth forms

Tree or shrub	Perennial (lives for more than two years) and usually woody	e.g. Acer, Viburnum
Forb or herb	Annual, biennial or perennial; herbaceous rather than woody	e.g. *Viola*, *Digitalis* (foxgloves), *Petunia*, *Zinnia*
Graminoid	Grass-like, with a single seed leaf and strap-like leaves	e.g. sedges, grasses, rushes
Geophyte (also called cryptophyte)	Having underground bulbs or rhizomes	e.g. *Allium* (onions)
Vine	Climbs or twines using tendrils, or backward-pointing thorns, prickles or spines	e.g. *Clematis* (twining), climbing roses (prickles)
Hydrophyte	Aquatic	e.g. *Nymphaea* (waterlilies)
Parasite	Takes its resources from a host	e.g. mistletoe
Epiphyte	Lives on other plants	*Tillandsia* (air plants), epiphytic orchids
Succulent	Stores water in leaves or stems	e.g. *Sempervivum* (houseleeks)
Ferns	Non-flowering, spore-bearing plants	e.g. *Asplenium*, *Dicksonia*

Plant growth

While some adaptations are clearly visible, such as the aerial roots of ivy clinging to a fence, others concern the way plants grow. Plant growth depends on photosynthesis, transpiration and respiration.

Most plants use a process called photosynthesis to capture light from the sun and carbon dioxide from the air and turn them into sugars for growth, while also releasing oxygen back into the air. To capture the carbon dioxide, the leaves contain holes called stomata, which allow the exchange of gases between the leaf and the atmosphere. The stomata also let out water vapour in the process of transpiration. To avoid losing too much water, plants can close their stomata, but this means that photosynthesis also stops – or does it? As we shall see in more detail later, some plants have evolved different types of photosynthesis to help them keep growing while reducing water loss.

Meanwhile, the sugars made through photosynthesis are converted back to usable energy for making new tissues through respiration (much as in humans), a process that requires the presence of oxygen. In waterlogged soil there may not be much oxygen, so plants that grow in those conditions have found ways around this low-oxygen problem.

This chapter introduces a number of terms associated with plant growth. They will be explained as they are used, but for convenience, opposite is a list of the main terms and their meanings. A full glossary can be found on page 216.

Salix

Term	Meaning
Photosynthesis	The conversion of carbon dioxide and water into sugars and oxygen, using light energy from the sun. There are three main types: C3 ('normal' photosynthesis) C4 (first stores carbon dioxide in a molecule with four carbons during the day) CAM (first stores carbon dioxide in a molecule with four carbons during the night). For more on photosynthesis, see pages 28 and 90.
Respiration	The release of energy to build tissues by converting sugars and oxygen into carbon dioxide and water.
Transpiration	The movement of water through a plant, driven by water molecules evaporating from the leaves (particularly through open pores called stomata).
Resources	What plants need to grow. This includes essential nutrients, such as nitrogen, phosphorus and potassium (referred to as NPK in fertilizers), as well as micronutrients and water. Plants also need carbon dioxide and oxygen for photosynthesis and respiration.
Enzymes	A chemical substance that helps reactions occur. RuBisCO, for example, is central to photosynthesis.
Hormones	Signalling molecules – just as in humans. These molecules are needed only in very small amounts, and changes in those amounts help plants to know when to change their behaviour. For example, an alteration in soil moisture prompts fluctuations in the hormone abscisic acid, which moves to the leaves and tells the stomata to close.
Pollination	The process by which a pollen grain from a male flower (or male part of the flower) reaches the egg cells in the ovary of the female flower (or female part of the flower). Often helped by wind or pollinating organisms, such as bees, butterflies, beetles, flies, birds or mammals.

Hidden superpowers
Below-ground organs

Standing among the perennials, one foot on the garden fork, we drive it into the soil and lever the roots from the ground. A survivor of the last dry period, this will be perfect for dividing and planting in other parts of the garden. But what's this? A tuber? Or a bulb? Is it a corm? Or maybe a rhizome?

Many plants that are able to tolerate extreme weather conditions form below-ground storage organs, either from their roots or from stems that remain in the soil.

Roots, rhizomes and underground stems

Below-ground storage roots are called tubers (or root tubers) and are formed of extra layers of cells that can store water or nutrients. Different parts of the root, or the whole root, become thickened in different species, and the amount of swelling depends on the conditions in which the plant is growing. Plants that form tubers include sweet potato (*Ipomoea batatas*; opposite), dahlias (*Dahlia* spp.) and cassava (*Manihot esculenta*).

Some roots have another superpower: the ability to contract, pulling the plant deeper into the soil. These contractile roots often occur together with other below-ground storage organs to keep the organ at the right depth. They work by pulling long cells, making them wider and shorter. The root tips (and the non-contractile roots) remain firmly in the soil, so the contraction (shortening) of the cells pulls the stem or bulb further into the soil. If hyacinths are grown in a glass jar, for example, the wrinkling that can be seen on some of the roots indicates that they are contractile. This feature helps plants in flood-prone areas to rebury themselves (see waterlilies, page 42, and *Symplocarpus*, page 70), or can produce more space for roots to form, as on the base of the yucca trunk (page 192).

Many plants have below-ground stems and even leaves, and these can also be used to store water or sugars. They include rhizomes, corms and bulbs – but what's the difference?

A rhizome is a swollen stem, and as such it has nodes (from which shoots form) and internodes (the stem between the nodes). These below-ground stems may grow leaves and roots if required. Gingers and some irises are examples of plants that produce rhizomes.

A corm is a short section of stem (sometimes a single internode) that swells, often giving the appearance of a bulb, but if cut open, it will be seen to have the cellular arrangement of a stem. A corm may also be surrounded by a protective tunic (leaf sheath) made of flattened leaf stalks. Examples of plants that produce corms are *Crocosmia*, *Liatris* and *Colchicum*.

Finally, a bulb is made up of swollen leaf bases (often called scales) arranged in layers – think of an onion. At the centre of the swollen leaves, protected from the elements, is the growing tip, which sends up new leaves and flowers. Aside from onions (*Allium*), other bulb-forming plants include *Hippeastrum* and tulips.

A different kind of underground storage organ is the potato (*Solanum tuberosum*). This is actually a storage organ that forms on stolons, which are a different type of stem. Stolons are more usually associated with above-ground runners, as in strawberries. In the potato, these above-ground runners are buried and form the tubers that become potatoes, and this is why a classic garden-scale way to grow potatoes involves building a tower of tyres. Starting inside one tyre, some of the shoots are buried in the compost (these form potatoes) and the remaining shoots grow upwards. Then another tyre is added with more compost, encouraging more stolons to swell, and so on throughout the growing season.

Ipomoea batatas

Plants that bury the lead

As the fork levers out the rhizomes of the plants to be divided, out falls a peanut! This one was probably buried by a squirrel, but peanuts (*Arachis hypogaea*; right) and some other plants do actually bury their seeds in the ground. This habit led to the peanut's other common name, groundnut (also used for Bambara groundnut, *Vigna subterranea*). Both are legumes that flower above ground, then bury the developing seed in the soil.

Another example is a common wildflower, ivy-leaved toadflax (*Cymbalaria muralis*). After the flowers have finished, the seeds develop on elongated stalks that look a little like tendrils. These stalks grow upwards at first, then find darkness in cracks in stone walls and bury the seeds deep within. This is how ivy-leaved toadflax eventually climbs walls without technically being a climber.

While burying seeds might make perfect sense for germination, some plants actually flower underground. An interesting example is a palm called *Pinanga subterranea*, which is native to Borneo and was discovered as late as 2023. It grows leaves from the soil (as do many palms), but produces flowers and fruit below the surface. How pollination occurs in these buried flowers is still a mystery.

As if that weren't weird enough, there are even plants that live their entire life below ground. A group of orchids in the genus *Rhizanthella* live underground and get everything they need through a relationship with fungi. These are examples of non-photosynthetic plants. The flowers are held just at soil level or beneath leaf litter, and, as with *Pinanga subterranea*, little is known about their pollination.

Arachis hypogaea

Plants that shirk dirt

While this section has focused on below-ground adaptations, some plants don't live in the ground at all. Some aquatic plants float free in or on the water. Water hyacinth (*Pontederia crassipes*; right) is so good at floating that its leaves can stick out well above the surface of the water. Below, roots and stolons remain free-hanging in the water. Away from its native range of tropical South America, these superpowers make water hyacinth more supervillain than superhero. An incredibly fast grower, producing new plants along the stolon and not anchored to the ground, it is a voracious weed in many parts of the world.

Some plants spend their entire lives in the tree canopy. These aerial plants, which are known as epiphytes, are made up of a huge diversity of species, from mosses and ferns to many types of flowering plant. The most popular orchid house plant, the moth orchid (page 142), is just one example.

While living on other plants might mean getting more light, epiphytes experience many other environmental challenges. To hold on to their position on trunks or branches, many exude glues from their roots, or wedge themselves into cracks in bark or the forks of branches. Nutrients are also a challenge, so some plants trap leaf litter in nest-like leaf arrangements. Others, such as tank bromeliads, hold pools of water in their leaves, attracting frogs or insects, which poop or die in the tank, releasing nutrients that the plant uses to grow. Like nutrients, rain might come in short pulses, and without soil to hold it, epiphytes need another way of keeping the water until they can take it up. A common solution is to grow sponges along their aerial roots, which hold moisture and can even condense water from the air. Air plants (page 144) have no roots at all and take in moisture through their leaves.

Pontederia crassipes

Superpowers on show

Above-ground organs

Mounds of dark green tumble over the retaining wall like hair on a child's dressing-up wig. Behind, fat green cacti contrast structurally with the soft appearance of *Casuarina glauca* ('Cousin It'). While they look very different, the two plants have something in common: they do not grow leaves. The green 'hair' is in fact stems, while the cactus's leaves have been reduced to tiny microscopic scales. Whereas *Casuarina* has shrunk its leaves out of existence, cacti have mutated them into spines.

Reducing leaf area while increasing the ability to store water is a common strategy for plants where water is limited. In combination with this, many plants have also evolved variations in photosynthesis to help them save water.

Leaves and stems

Above ground, plants' leaves and stems are exposed to extremes of temperature, bright light and drying winds. These conditions drive water evaporation from the leaves, so

Aeonium arboreum

when soil moisture is unavailable (whether frozen or absent), plants need ways of reducing water loss.

One strategy is to store water in leaves or stems, quickly soaking up moisture from even the shortest downpour and allowing the plant's cells to swell with the life-giving liquid. Succulence can occur in any plant organ, including those below ground, and is often part of the benefit of rhizomes and tubers. Above ground, we need only look at the African baobab (*Adansonia digitata*) or Queensland bottle tree (*Brachychiton rupestris*) for impressive examples of water being stored in trunks. Cacti also store water in their swollen stems, while the plants that we usually refer to as succulents are recognized for their ability to store water in their fleshy leaves. These include the juicy leaves of *Aloe vera*, *Sempervivum* (page 130) and many species of *Euphorbia* (page 90).

Cattleya

Succulents, then, are – not surprisingly – succulent. Less obviously, many rainforest epiphytes also use succulence to store water. Rain may be frequent, but life in the canopy can be tough. As soon as the rain stops, the water runs off and the plant is effectively in a drought. To avoid this, many epiphytes have fleshy leaves and thick aerial roots. The moth orchids (*Phalaenopsis* spp.) mentioned on page 25 are a classic example. Others, such as *Cattleya* spp. (above right), even have a fleshy rhizome that can store water. This is often called a pseudobulb (false bulb) because, although it looks like a bulb, it is not made from leaf bases (see page 22).

Storing water in succulent stems and roots means the plant can access water during dry periods. Minimizing the size of leaves or removing them altogether, meanwhile, reduces moisture loss because there is much less surface area from which the water can evaporate. But if there are fewer pores (stomata) for releasing water, less carbon dioxide is able to enter the plant for making sugars. That brings us to the plants that have modified the typical photosynthetic pathway to be more efficient.

Capturing carbon dioxide or wasting water?

What we think of as 'normal' photosynthesis involves carbon dioxide entering a leaf through stomata and being converted to a three-carbon molecule by an enzyme called RuBisCO (short for the not-very-punchy Ribulose-1,5-bisphosphate carboxylase/oxygenase). The three-carbon molecule is the origin of the name 'C3', which is used for this type of photosynthesis. Meanwhile, light from the sun is converted into energy that drives the many steps involved in creating sugars. While this 'normal' process costs the least energy of the three main types of photosynthesis (see page 21), it results in the loss of up to 95 per cent of the water that is taken up from the roots. It also loses back into the atmosphere any carbon dioxide that has not been fixed in photosynthesis.

The goal of plants from dry environments, then, is to make the most of every molecule of carbon dioxide and reduce the time the stomata are open. This is exactly what the two main alternative photosynthesis strategies do.

Instead of first fixing carbon dioxide into a three-carbon molecule, C4 photosynthesis uses a four-carbon molecule. That molecule is moved to cells near the veins, where it releases carbon dioxide, which is then converted into the three-carbon molecule of C3 photosynthesis. The advantage is that more of the carbon dioxide that enters the leaf can be captured, allowing growth to occur even when the stomata are open for a shorter period. Closing the stomata reduces water loss from the leaf. Plants that use C4 photosynthesis include maize (corn; *Zea mays*) and *Atriplex* spp. (page 191).

Carbon-capture at night

C4 photosynthesis happens during the day, when there is a stream of light energy to fuel the reactions. Another group of plants has found a way to capture carbon dioxide at night, when less water will be lost from the stomata, then convert it to sugars during the day, while the stomata are closed. This process is called CAM photosynthesis (short for Crassulacean Acid Metabolism) and was first identified in succulents belonging to the family Crassulaceae.

Very similar to C4 photosynthesis, CAM fixes carbon dioxide into a four-carbon molecule (which is an acid) and stores that in the vacuole (the fluid-filled cavity in the cell). This occurs during the night, when the stomata are open. Then, as the sun rises and with it the temperature, the stomata are closed to prevent water loss. Carbon dioxide is released from the stored acid and provided to the RuBisCO enzyme to make the three-carbon molecule that leads to the chain of reactions that makes sugars. However, there is a limit to how much of the four-carbon acid can be stored, so these plants are often slow growers. Some plants (including many epiphytes) can switch between C3 and CAM photosynthesis, using the more energy-efficient C3 process when water is not limited.

Other powers

These water-saving strategies are not the only superpowers that enable plants to survive in harsh conditions. We'll discover more in later chapters.

Agave americana

Strutting their stuff
Reproductive superpowers

None of the superpowers mentioned so far is any good at all if the plants are unable to pass it to their offspring. For this to happen in flowering plants, pollen from the male part of the flower (called the anther) must be passed to the female part of the flower (the stigma). The pollen then grows a long tube (called a pollen tube) that reaches into the ovary. The two haploid (carrying half the genes) male gametes from the pollen travel down the tube into the ovary, where they fertilize the female egg cells (which are also haploid gametes). This produces an embryo containing a full set of genes and leads to the production of the seed.

If we describe it like that, it doesn't sound all that different from making a baby. The problem is that plants can't get up and go on a date. Instead, they need a go-between to transfer the pollen from the anther of one flower to the stigma of another flower (or, in some cases, pollinate themselves). This could be passive, such as blowing in the wind, or active, by attracting birds, mammals or insects using nectar and other rewards.

To attract their go-between, some plants put on quite the blowsy show, with bright colours and even some hues that humans can see only by using an ultraviolet lamp (also called a blacklight). Bees can see these colours easily, however, and flowers use them as beacons to show them the way. Other flowers attract pollinators using scent – sometimes not to our liking, as with carrion flowers, which release a smell like that of rotting meat to attract pollinating flies. Yet others use heat to help the scent dissipate further into the atmosphere, or – as hellebores (page 104) and some other winter-flowering plants do – to reward their insect visitors with a warm roost while they collect the pollen.

While some flowers can be pollinated by a wide range of animal species, others have evolved very specific partners. An example is the Joshua tree (page 192), which is pollinated by two closely related species of moth.

Most flowers provide a reward of some sort, but others are sneaky. The bee orchid (*Ophrys apifera*; opposite) is an example of a cheeky flower. Its flowers mimic the female of a particular bee species (*Eucera longicornis*) in both scent and appearance. The male bees find the flowers irresistible and try to mate with them, during which the flower plonks a packet of pollen on the bee. When the bee flies off frustrated, it takes the pollen with it – straight to another irresistible flower.

There are many, many fascinating and bizarre ways in which flowers ensure pollination. But the purpose is always to produce seeds that will grow into new plants carrying the adaptations of their parents. Since many of these seeds are also edible, have a quick look at Chapter 10 to learn about the different types.

Ophrys apifera

Feeling hormonal

In the garden shed, with pots of gritty soil to one side and shrubby branches to the other, short cuttings are taken from the branches. The lower leaves are removed and the base of each cutting is dipped into a container of 'rooting gel'. But what is this magical substance that helps roots to grow?

Like people and animals, plants have hormones that help them to grow and respond to their environment. One of those hormones, auxin, is the most common active ingredient in rooting gels and powders. Auxin is produced in shoots and moves down the stems towards the roots. When the roots are removed (as when we make a cutting), auxin builds up at the cut end and instructs the cells to divide. These dividing cells then decide to make a root to replace the lost ones. Willow (*Salix* spp.; opposite) is very high in auxins all the time, which is why it produces a new plant so readily if we cut a stem and stick it straight into the ground.

Willow is also high in another hormone that is very important for responding to wounds and defending against attack by herbivores: salicylic acid. This chemical, which is also high in meadowsweet (page 63), is the active ingredient in aspirin. It works as an anti-inflammatory in people, while in plants it boosts the immune system to limit the damage caused by pathogens. In plants, this is called systemic acquired resistance. One variation of salicylic acid is given off by plants as a gas, acting as an early-warning system. Nearby plants take in the gas through their stomata and turn it back into regular salicylic acid, which switches on their defences.

The last plant hormone to introduce here is ethylene. This is a gas most commonly known for ripening fruit. For example, bananas produce a lot of ethylene, and when put in a bag with unripe tomatoes, they will speed up the ripening. But ethylene is also very important for plant responses to environmental conditions. Plants make ethylene all the time, but as a gas, it wafts into the air or pores in the soil, preventing it from building up. During a flood, the water prevents the ethylene from wafting away, and instead it builds up. This build-up tells the plant to develop features that will help it to breathe under water. One of these is the production of tubes called aerenchyma in the stems and roots (see Chapters 2 and 3). These tubes act as straws, pulling air through the stems and roots to actively growing tissues, thus keeping them alive until the flood passes.

There are numerous other hormones that help plants to grow and respond to their environment. Abscisic acid helps to trigger the closing of stomata; jasmonic acid is instrumental in wound response and, in gaseous form, signals to nearby plants; cytokinins work with auxins to control cell division and growth; and gibberellins are typically known for influencing cell growth.

Salix

Gardens as ecosystems

Although in this book we focus on plants, it is impossible to ignore the wider life forms that support and are supported by plants.

We all love the sound of bees humming, birds chirping and frogs croaking away in each part of the garden. We love to see butterflies and dragonflies fluttering around and to find ladybirds (ladybugs) among the foliage. But these are just the tip of the biodiversity iceberg. At night moths fly about, attracted by night-scented (often pale-coloured) flowers, while other essential pollinators in the ecosystem include beetles, wasps, hoverflies and flies. Bats dip and dart at dusk, helping to control the number of flying insects (including mosquitoes), while hedgehogs and lizards roam in search of slugs and snails.

This macrofauna, although great to watch, is not the whole story in the garden ecosystem. Mostly too small for us to see, micro-organisms are central to a healthy garden. In the soil, fungi called mycorrhizae make associations with the roots of plants, feeding the plants nutrients foraged from the soil, while the plants feed the fungi with sugars to help them grow and forage more. Other fungi are important in the cycling of nutrients, breaking down sticks, logs and dead leaves and releasing nutrients into the soil for the plants to soak up.

Legumes make little nodules on their roots as a home for special bacteria called rhizobia. These bacteria are able to take nitrogen from the air (the Earth's atmosphere is 78 per cent nitrogen) and give it to the plant to use. As with mycorrhizal fungi, the plant then feeds the rhizobia sugars from photosynthesis. This may be a familiar association, but did you know that legumes are not the only plants to use microbes in this way? Alder (page 72) and *Casuarina* (see page 26), along with many others, also use microbes to 'fix' nitrogen from the air, and some epiphytes seem to be home to nitrogen-fixing bacteria called cyanobacteria.

Other micro-organisms called yeasts are adopted in the flowers of hellebores (page 104) to create heat that attracts pollinators in the cold of winter, when these plants flower. The plants provide the yeasts with sugar from nectar, which drives very fast growth in the yeast, producing warmth.

The more we look at this microbial world, the more we learn about the beneficial role of microbes in nutrition and as defence against disease, both inside the plant's tissues and on its surfaces. And that's just the interaction with plants!

Diversity is key to supporting healthy garden ecosystems at all scales. Simply put, more plant species can support more wildlife. To host the most plants, gardens need a diversity of niches (habitats that support different plants): low places for rain runoff and water-loving plants; higher, well-draining places for plants that don't like to have wet feet; rockeries and stumperies, each supporting different kinds of plant life and providing habitat for other organisms. And don't forget to feed the larval stages of your favourite lepidopterous friends. If we want fluttering butterflies and moths, we must feed the caterpillars!

Nelumbo

Chapter 2

Water Submergence: Aquatic Plants and their Adaptations

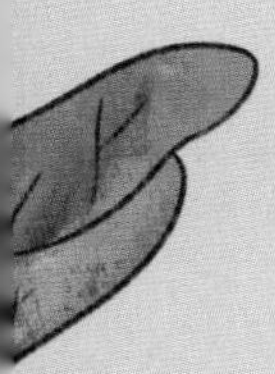

Introduction
Water Submergence

A pond. Ripples radiate gently from reeds standing tall in the water. A dragonfly hovers, dips towards the surface, then darts away. But such a calming scene is not just good for our well-being; with warmer, drier weather, garden ponds are also becoming increasingly important for supporting wildlife.

The plants that live in this watery world provide shelter, food and oxygen for the wildlife that depend on the pond. To maximize their value, ponds should ideally include a combination of plants that float, some that grow with their roots in the water and their foliage above (called emergent), and some that live in the watery depths, photosynthesizing and therefore increasing the oxygen in the water.

Water plants (also called aquatic plants) live in a very different set of environmental conditions from the terrestrial plants that we explore elsewhere in this book. In the water, light, carbon dioxide and oxygen are all reduced. The pressure under which the plants grow is higher, but buoyancy counters the effect of gravity. Since these changes are so important for pond plants, let's explore them one by one.

Typha

Seeing the light

We have already seen that plants need light – particularly blue and red light – for photosynthesis (see page 20). However, water filters light, changing both the colour and the amount that is available to underwater life. Red light is the first to be filtered out by water, and the deeper we go, the less red light is available for photosynthesis. Garden ponds are usually not deep enough for this to become a problem, but in the ocean the limited red light is the reason some seaweeds appear red. The light-harvesting pigments for photosynthesis in these algae maximize absorption of blue light and have no need of the absent red light, so the fronds reflect this colour. This reflected red light is what we see when we look at the seaweed.

Clear pond water can still filter out up to 30 per cent of red light within the first metre (3ft). If particles are present, such as stirred-up sediment, algae or other micro-organisms living in the water, the amount of red light reaching the bottom of a garden pond can be very limited, making photosynthesis difficult for our beloved pond plants.

Water plants have come up with some amazing ways to deal with this reduced light. Some avoid the problem altogether by floating their leaves to the surface or standing up above the water. Many of those that live in the water itself limit other restrictions to light, such as having thin leaves so as to reduce the distance light travels within the leaf (and reducing the chance of the light bouncing back out of the leaf), or having less of the waxy protective layer called the cuticle, which coats the leaves of terrestrial plants. In terrestrial environments this waxy cuticle is important for reducing water loss and reflecting excess light – but in a watery environment there is no need to reduce water loss, and light is limited rather than being in excess.

Cyperus longus

Out of thin air

It's much harder for gases to mix with water than with air – approximately 10,000 times harder, in fact. This is part of the reason humans can't breathe underwater, and why fish and some other aquatic creatures have specialized gills to capture as much oxygen as possible. For plants, it means that underwater leaves have less access to carbon dioxide for making sugars, while roots have less access to the oxygen they need to turn the sugars into tissues through respiration.

One way in which some plants overcome this is by having their own equivalent of diving gas tanks. Instead of keeping the gas inside a container, they hold it against the outer side of the leaf surface. This is particularly useful for plants that experience changes in water level, such as reeds, which are normally emergent but can become submerged after a large storm or heavy rain. This means that photosynthesis can continue in the submerged leaves until the layer of air runs out or the water level drops.

While a tank of air is helpful for a limited time (just as in diving), a snorkel can allow breathing indefinitely. Many plants build their own snorkels within their tissues. This allows emergent or floating leaves to exchange gas with the atmosphere, while their roots remain below the surface (even anchored to the substrate). The stems of these plants develop long, hollow tubes called aerenchyma that allow gases to move into the submerged roots from leaves above the surface of the water. In this way, the roots and rhizomes can keep growing using oxygen from the surface.

Nymphaea

Under pressure

One aspect of pressure is the amount of water weighing down from above. The deeper we go, the more water is above us, weighing down on us. Another aspect is buoyancy. This is the force with which the water pushes an object upwards. When we go for a swim in a pool, the water pushes our body up, helping us to float. We tend to float because we are less dense than water, while a rock is denser than water so will sink. We take advantage of the buoyancy and consistent pressure of the water that envelops us at shallow depths for rehabilitation from some injuries. These features can also be exploited by plants.

The floating leaves of waterlilies and lotuses are filled with air pockets, while those that remain submerged contain less air, wafting with the current. Meanwhile, the increased pressure provides support around the plant, reducing the need for a 'skeleton'. In land plants, structural support tends to come from a compound called lignin, which is what makes wood strong and allows trees to stand so tall. In aquatic plants the water provides some of this support, so many such plants produce fewer of these structural compounds, making them soft and floppy so that they wrap themselves around fingers scooping them from the water.

The lifeblood

While some aspects of a water world may seem challenging, one thing is not a limiting factor: water. In a changing climate with more extremes, this lifeblood of all ecosystems will become ever more valuable. Capturing water during heavy storms in ponds or tanks and creating favourable habitats for wildlife during times of little rain will make a real difference to local biodiversity. From the smallest container water gardens to large dug-in ponds, these oases provide myriad benefits: a quick drink for pollinators; breeding grounds for organisms that have an aquatic life stage, such as damselflies and frogs; and drinking water for larger organisms, such as hedgehogs, birds and lizards, some of which also depend for food on the insect life the pond supports.

In this chapter we'll explore how some of the common plants we find in garden ponds use these features to great effect. Not just surviving but also thriving, these species help us to create the wildlife havens that are increasingly important in a changing climate.

Cyperus longus

Nymphaea species
Waterlilies

The first plant I think of when someone mentions ponds is always the delicate waterlily. The flat green leaf pads create effortless floating bridges, accented by their regal flowers, which reach above the still water, enticing and perfect. These beautiful plants are fascinating for their ability to remain firmly fastened to the waterlogged soil at the bottom of the pond, grow through the water and access the atmosphere above it. They have a number of interesting features that help them to do this.

Starting in the murky depths, waterlilies grow from tubers and rhizomes (storage organs similar to sweet potatoes). The roots that emerge include roots for nutrient uptake from the fertile debris at the bottom of the pond, but waterlilies also grow a specialized type called contractile roots. These are thicker than standard roots and can expand or contract to pull the tuber deeper into the substrate, anchoring it firmly at the ideal depth for growth to be maintained.

The shoots also emerge from the tubers or rhizomes, the stored energy fuelling the fast growth as the leaf buds race for the light and air at the surface. If you have a waterlily growing in a pond and count the leaves each day (especially in the spring), you'll see the buds reaching up from the bottom one day and fully unfurled at the surface the next. This ability helps them to adjust to the deeper water after heavy rain.

Waterlilies use air-filled chambers in the leaf to float the buds to the surface, and, once they are unfurled, these air-filled spaces keep the leaf flat on top of the water. If we cut a leaf and looked at it under a microscope, we would see these big air spaces filling the lower cell layers (the side that is in contact with the water), while the top cell layers, which have unrestricted access to sunlight and where photosynthesis takes place, are packed with many more cells filled with green chlorophyll.

Since only one side of the leaf has access to the air – the upper surface – this is where the pores for gas exchange (stomata) are also found. These stomata allow in carbon dioxide for making sugars through photosynthesis, but they are also connected to the air spaces throughout the leaf and stem, and act as straws to draw oxygen all the way to the rhizomes and roots. This is important because plants need oxygen to convert the sugar into growth using respiration, but the waterlogged soils in ponds are usually very low in oxygen. The air tubes connect the roots with fresh air in much the same way as a snorkel.

As if that weren't amazing enough, some waterlily flowers use respiration to create their own heat (see *Symplocarpus*, page 70), wafting their scent into the early evening air to attract pollinating beetles.

All in all, while these elegant plants may have been among the earliest flowering plants (emerging about 140 million years ago), their fascinating adaptations make them some of the most interesting of all.

Nelumbo species
Lotuses

Beads of water roll around the upturned leaf of a lotus, the tall, proud flowers of which are similar in elegance and beauty to the waterlily's (page 42). Despite their similarities, as dicots lotuses are more closely related to banksias and proteas (family Proteaceae) or plane trees (Platanaceae) than to waterlilies, which are monocots. In fact, the similarities between these plants constitute an example of convergent evolution: when two unrelated organisms adopt similar features to adapt to similar environmental conditions.

Lotus leaves are remarkably well adapted to keeping their surface clear of water and dirt, and those beads of water rolling around the surface are the clue to how they do it. A bead of water forms when the individual water molecules prefer to hold on to one another than to cling to another surface. One of the features that decides if a water bead will form is the roughness of the surface the water is touching, in particular the roughness at the tiny scale that a molecule of water experiences. Lotus leaves are covered in microscopic bumps, creating micro-valleys in which air pockets form. This means the water molecules are in more contact with one another than with the leaf surface. This leads to the formation of a water bead, which rolls around the leaf collecting dirt before eventually rolling off it entirely.

This property is called superhydrophobicity. If we break the word down, we find the Latin *super* (very), Greek *hydro* (water) and Greek *phobos* (fear). Basically, think of a surface that is extremely 'afraid' of water. Researchers have been inspired by the water-repellent nature of lotus leaves, and similar microscopic surface properties are now being used to formulate household paints that repel dirt and water and reduce the need for harsh cleaning chemicals.

Lotus flowers, meanwhile, can produce their own heat. They are not alone among plants in having this ability (see *Symplocarpus*, page 70, and *Helleborus*, page 104), but they are unusual in that they adjust the heat they produce based on environmental conditions. Lotus flowers, in common with those of most flowering plants, have petals, stamens (where the pollen is produced) and receptacles (like a womb, where the seeds will develop after pollination). Each of these three flower parts can produce heat. Heat is generated through respiration – burning carbohydrates to release energy – and the amount varies according to the air temperature. When that drops, heat production increases until the air warms up again. This means the flower keeps a steady temperature during pollination.

This is a warm reward for pollinating insects, but it has also been shown that the flowers need warmth for successful seed initiation even when they are pollinated manually. After seed development begins, heat production is no longer needed.

Callitriche stagnalis
Common water starwort

Like rolling green hills, undulating drifts of common water starwort rise and fall below the surface, punctuated by emergent branches with larger leaves that exaggerate the scale, making those hilltops seem much closer to my peering face. Native to Europe and North Africa, this fast-growing aquatic plant has become a weed on other continents through accidental release into drains (so, as with every plant species, always check for suitability in your area). Its success is down to two factors: high seed production through self-pollination, and the ability to grow new plants from small segments.

A fascinating aspect of many aquatic plants is that leaves growing above the water's surface can be very different from those growing below it. Underwater leaves are often just two cell layers thick, with every cell containing chloroplasts (chlorophyll-containing particles) for photosynthesis. By contrast, above-water leaves produce additional cell layers, and only the outer layers contain chloroplasts. Being above water increases access to carbon dioxide and light, but too much light can damage the photosynthetic equipment. To avoid this, above-water leaves have a thicker cuticle. But this also restricts carbon dioxide entering the leaves, so more stomata are produced to let it in. Without the risk of too much light, underwater leaves have no need for a cuticle and instead absorb carbon dioxide directly across every cell.

Having two growth forms in this way is called dimorphism (from the Greek *di* = two and *morphe* = form), and water starwort is one of these dimorphic species. The appearance of underwater rolling hills is created by the dense mat of small leaves, while the emergent leaves are larger and thicker. If you have a pond or live near a waterway, follow the emergent leaves to their submerged neighbours and see how many aquatic plants have shape-shifting superpowers.

Myriophyllum spicatum
Spiked water milfoil

One of my favourite oxygenating plants in my own small tub pond in the UK is spiked water milfoil. The feathery leaves give this plant a delicate appearance that contrasts nicely with the emergent iris growing next to it.

An interesting thing about *Myriophyllum spicatum* (and some other water plants, including *Ranunculus aquatilis*; page 48) is that it can take up bicarbonate from the water and use it as a source of carbon dioxide for photosynthesis. You may be familiar with bicarbonate from bicarbonate of soda/baking soda (sodium bicarbonate), a raising agent in many recipes. In baking soda, carbon dioxide is released from the bicarbonate into the cake mix to make it fluff up as it bakes.

In plants, once it reaches the leaves, the carbon dioxide can be released with the help of enzymes or through acidification. You can observe the acidification process in your kitchen by squeezing a lemon over some bicarbonate of soda, which will fizz, releasing bubbles of carbon dioxide. Once the carbon dioxide has been released from the bicarbonate within the leaf, it can be used for photosynthesis.

But where does the bicarbonate in water come from? As water runs over rocks, especially limestone or dolomite, it picks up bicarbonate. This is because limestone and dolomite are made of calcium carbonate or magnesium carbonate. Stalagmites or stalactites are created by water dissolving the bicarbonate and leaving it behind when the water evaporates. Once bicarbonate is dissolved in water and joins waterways, it can be used by aquatic plants, such as spiked water milfoil, for photosynthesis.

As with any plant, be very careful with species names when ordering. The closely related *M. aquaticum* is invasive in some places, including the UK.

Ranunculus aquatilis

Common water crowfoot

Forming floating green mats peppered with white buttercup-like flowers with yellow centres, the attractive water plant *Ranunculus aquatilis* is native across Europe, North America and North Africa.

Like *Callitriche stagnalis* (page 46), *R. aquatilis* has dimorphic leaves. In this case, the difference in leaf shape between the submerged and floating forms is dramatic. The floating leaves loosely resemble those of clover, with deeply divided teeth, while the submerged leaves are thread-like, almost feathery, like those of *Myriophyllum* spp. (page 47). This dramatic dimorphism of leaf type is called heterophylly (from the Greek *heteros* = different and *phyllon* = leaf).

Having large floating leaves can maximize the interception of light, and the many stomata allow in the more available carbon dioxide from the atmosphere. Meanwhile, the threadlike nature of the submerged leaves minimizes the distance across which gases have to diffuse, and reduces the risk of damage in a current by presenting less resistance to the flow of water.

It's fascinating how the same plant can produce one leaf shape above water and a completely different one when submerged. Like people, plants have genes and hormones, and it is the interaction between genetic controls and hormone signalling that determines leaf shape. In submerged *R. aquatilis*, a reduction in the hormone gibberellin triggers the genetic pathway that leads to divided leaves. While gibberellin drives leaf shape in many aquatic plants, other species respond to changes in another hormone: ethylene. Ethylene (page 32) is a gas that builds up in submerged plants and is also important for triggering the gas tubes that occur in waterlilies (page 42) and other plants.

Nasturtium officinale
Watercress

Very familiar from the supermarket shelves, *Nasturtium officinale* is an edible leafy vegetable that is native across Europe and Asia. Vigorous in aquatic habitats thanks to its fast growth, hollow stems for gas exchange, and ability to regrow from small stem segments, watercress can become a weed outside its native range.

Typically of members of the mustard family (Brassicaceae), watercress has a peppery flavour that livens up salads. But in its natural habitat the chemical behind this flavour is a defence against herbivores, such as caddis flies and snails. As these creatures chew the green tissues, two chemicals (myrosinase and glucosinolates) are forced together. The reaction between them creates a range of oils that are toxic to many herbivores and that also make the peppery flavour.

As an interesting aside, the group of garden plants commonly called nasturtiums are not only a different genus (*Tropaeolum*), but also in a different family (Tropaeolaceae). However, they get their common name from the oil they produce, which is also found in watercress.

Wild watercress is certainly edible, but it's important to be very cautious about eating it, since parasites from animal faeces can live in its tissues. The watercress sold in supermarkets is typically grown hydroponically (without soil, in controlled conditions), so is free of this risk. However, as a pond plant, watercress is a pretty addition with its delicate white flowers and lush foliage.

Mentha aquatica
Water mint

Blurring the boundary between open water and embankment, water mint bobs gently in the wake of a gliding swan. Stalks bearing pale pink pompoms, buzzing with the sound of bees, stand tall above the aromatic green foliage. On closer inspection, the pompoms turn out to be made up of many small flowers typical of the mint family (Lamiaceae). But below this apparently terrestrial growth form is another story.

Stems meander through the water to the rhizomes embedded in the murky riverbed, and emerging from these stems are wispy white threads. These are actually roots, and they are key to the way *Mentha aquatica* thrives in its watery habitat. They constitute one of several types that are known as adventitious roots.

In a botanical sense, 'adventitious' simply means appearing in an abnormal or unusual place, and here it means roots forming from stems. I find this a very human-centric definition, however, since the more we look at plants, the more examples of these stem-formed roots emerge. Even more interesting is the role these root types play in plant adaptations to different habitats. We will see several more examples later in the book.

In water mint, these stem roots allow access to oxygen for growth and respiration. As well as being closer to the water surface, their position on the stem connects them with the aerenchyma straws (see page 40) that allow gas movement from the leaves above the surface to the underwater tissues.

With access to oxygen and nutrients in the water, these roots are well placed to supplement the growth needs of the plant. But this isn't the end of their role. If a stem is broken off, such as during a flood, the adventitious roots are able to maintain growth until the free-floating cutting finds purchase by an embankment or among reeds, where a new root system can develop in the substrate. This is a form of clonal reproduction, similar to taking stem cuttings from shrubs, and it results in the development of a whole new plant that is genetically identical to the parent. Clonal reproduction is a very successful strategy in certain habitats, including waterways, highlighting the need to check whether a plant might be locally invasive before adding it to a garden pond.

As these meandering square-sided mint stems grow through the water, they connect the roots anchored in the substrate with oxygen. Also filled with aerenchyma, the subterranean roots form thickened rhizomes that store resources through the winter. This food store enables fast growth as the weather warms, and supplies new runners that grow through the substrate and water towards lighter areas, where new shoots emerge. In turn, new wispy white root threads emerge, wafting in the water.

Typha species
Bulrushes

A familiar sight along waterways globally, bulrushes stand tall, their brown seed heads holding high the feather-light seeds for the wind to disperse. While we may be most familiar with the larger varieties, which stand as tall as 3m (10ft), there are some delightful smaller varieties, including *Typha minima*, which is suitable for even the smallest pond garden with its small pompom-like seed heads and fine green foliage.

Along with its hollow stems, another feature that makes bulrushes very tolerant of changing water levels is the leaves' ability to hold gases along the surface, like divers' gas tanks. This air layer means that photosynthesis can continue below the water surface until the 'tanks' run out. Similarly to the way lotus leaves (page 45) repel water using microscopic roughness on the surface, the microscopic roughness of bulrush leaves means that a layer of air is trapped around the leaf when the water level rises. The stomata continue to take in carbon dioxide for photosynthesis and release oxygen into the air layer. When this layer was removed in a recent study, photosynthesis was reduced to one sixth the rate that is possible with the air layer intact.

As if that weren't enough, many parts of bulrushes are edible, from the rhizomes and young shoots to the seeds, which can be used to make flour. It's a fascinating, attractive and useful addition to a garden of any size.

Iris ensata
Water iris

With their bright splashes of purple moving in the breeze, water irises could be mistaken for their terrestrial relatives. However, these emergent plants grow well in their watery environment and can adapt to periods of complete submergence.

Water irises produce aerenchyma (air straws) to draw oxygen into their submerged tissues for use in respiration, to release energy from sugars for building tissues. However, sometimes these oxygen straws are not enough to overcome the low oxygen conditions in the substrate. In such conditions, plants resort to an alternative pathway for making energy: anaerobic respiration (without oxygen), which involves making the alcohol ethanol from sugars. Yellow flag iris (*Iris pseudacorus*), for example, is very good at switching to this pathway when oxygen is limiting, to maintain metabolism until adequate oxygen returns.

While the ethanol pathway is able to maintain survival, it produces about one-fifteenth the amount of energy that regular oxygenated respiration produces. That is why most water plants make straws to suck oxygen to submerged tissues, rather than using anaerobic respiration.

Water irises really are superplants. These aquatic species are able to tolerate dry periods, during which the starchy rhizome remains dormant, waiting patiently for water to return before bursting forth, fuelled by its stored sugars. There are many species and varieties of these beautiful, adaptable plants, in many colours, so there is sure to be one for every water garden.

What makes a good wildlife pond?

There is certainly no one-size-fits-all answer to this question. Even small ponds provide value to wildlife, and are pleasing for people. Recent studies have shown that the diversity of species visiting a garden pond is similar to that visiting larger urban lakes. If we convert this so that we are comparing the same surface area of water (considering each water body by relative surface area), it has been shown that small ponds attract more wildlife visitors per unit area than large bodies of water.

But who might come to visit? Of course, this will vary depending on the site of your water garden and where in the world you are. If you have only a balcony several floors up, your visitors will probably need to fly. That means you could expect birds, pollinating insects and those with both aquatic and aerial life stages, such as damselflies. If you have a container pond in your garden, you might also find that amphibians make it their home, and if your pond is dug into the ground, larger wildlife might drop by. The species within each category will depend on your part of the world, and it's worth looking up your local wildlife groups to find out who might make your pond a home.

When making a wildlife pond, remember always to check which plant species could become invasive in your region. Every pond is connected to the wider environment, whether obviously or not, and a flood or even just cleaning it out could lead to escapees.

Here are the other important aspects to consider.

Size

If you live in colder climes, a pond or container that is deeper than 60cm (24in) at least somewhere in its area will be less likely to freeze completely in the winter.

Sides

Gradually sloping sides are ideal for letting wildlife in and out of your pond. If you are planning to use a container with straight sides, place a ramp on the outside and inside to allow access and avoid trapping any creatures in the water.

Location

With a larger pond, it's ideal to have some areas in the shade and some in full sun. If you are making a container water garden or have limited space, choose the plants that are best suited to the conditions you have.

Plant types

Always check the regulations in your area and choose plants that will not become invasive. Native species are usually a safe choice, but there are many non-natives that will also provide value and that may be well adapted to your local changing climate.

Emergent Plants that are rooted in the substrate and grow through the water to stand tall above the surface. These are particularly important for dragonflies and damselflies, the larvae of which must climb out of the water when they are ready to turn into adults.

Floating plants Plants that float on the surface. They create shelter for organisms that live in the water, provide platforms for insects to rest on and from which they can access the water, and offer shade in the summer, reducing the amount of oxygen-guzzling algae growing in the pond.

Marginals Plants that like to have their feet wet and can tolerate both damp ground and shallow water (Chapter 3 showcases some of these). Choosing pollinator-friendly plants for the edges of your pond will enhance the wildlife value of your water garden.

Oxygenators Plants that live in the water. They photosynthesize underwater, breathing out oxygen that is essential for a healthy pond. They also provide shelter for aquatic organisms.

Lysichiton americanus

Chapter 3

Wet Feet: Adaptations for Living on the Fringes

Introduction
Wet Feet

Midsummer, and despite the dry weather there is a green patch where cowslips and meadowsweet thrive. This damp dip in the landscape becomes a submerged waterway after the heavy downpours that have become more common in recent years. Is that a frog sitting on the edge?

Plants adapted to switches from one extreme to another will be central to climate-smart plant selections. One type to consider is those that can alternate between flooded and drier, more 'normal' conditions. In their native ranges, these amphibious plants have had to switch between dry and submerged conditions, finding ways to survive or escape the wet, low-oxygen conditions and readapt to drier conditions once the flood passes. In nature this sequential wetting and drying occur most frequently at the edges of waterways, such as creeks and rivers, and in swamplands, so it is not surprising that many of the plants in this chapter are native to such habitats.

Acorus calamus

Adaptable traits for wet and dry

One trait that brings an advantage in many habitats is the presence of underground energy storage, such as rhizomes. We will come across many such plants, which are known as geophytes, in this book. Having a stash of energy below ground means that once the stressful condition has passed, new shoots and roots can grow very quickly. Many amphibious plants have rhizomes to help them survive.

While many aquatic plants have aerenchyma (see page 40) as part of their normal development, plants that are temporarily flooded may form these air straws only when they are needed. We have explored their benefit for transporting oxygen, but having aerenchyma also saves energy, since cells are removed to make the air spaces. The energy that would normally be needed to keep those cells alive can be used elsewhere, which is an advantage whenever there is a resource limitation. During drought, for example, having fewer cells means that the stomata can be closed for longer, reducing water loss. However, while aerenchyma are great under stressful growth conditions, they do reduce the capacity to take up nutrients and water, and this is why they are present only in adaptable plants.

Alnus

Two more adaptations that help to bring in oxygen during flooding are cork phellem (a spongy, bark-like layer) and lenticels (pores). Cork phellem is the tissue in cork oak (*Quercus suber*), harvested to make corks. Some flood-tolerant species, such as *Lythrum* spp. (page 66), respond to submergence by producing a layer of phellem that creates pockets of oxygen for respiration. Lenticels, meanwhile, are stem pores that link the outside air to the phellem or aerenchyma. They are common in trees that undergo periodic flooding, such as alder (right and page 72) and mangroves.

Once oxygen is inside the plant, it's important that it doesn't leak back out, so plants accumulate a layer of the fatty polymer suberin, which creates a barrier to prevent air from moving backwards out of the cells. Suberin, which is made up of fatty units called monomers, repels water and is airtight. In plants it occurs in almost all roots, just outside the vascular bundles (where the water-carrying vessels are). Since plants already have this built-in barrier, it's not surprising that in flood- and drought-prone habitats, plants have evolved to create a second layer of suberin on the outside of the tissues that contain aerenchyma. In flooded conditions, this outer coat helps to keep in the oxygen so that it can reach the submerged tissues.

Energy in thin air: respiration without oxygen

When oxygen is present, aerobic respiration burns sugars to release energy in the form of electrons, which can be used to make new tissues. This process involves many steps, but one step uses pyruvate and acts rather like a junction in the step chain. The goal is to catch energy (in this case in the form of released electrons) and pass it step by step down a chain. At the end of the chain, the energy can be used for building new tissues, among other processes.

When oxygen is present, the electron catcher is oxygen. But when oxygen is low, a different catcher must be used. Often the first alternative road taken at the pyruvate junction is lactic acid – just as in our muscles, when they can't get oxygen quickly enough during exercise. However, too much lactic acid can damage the cell processes, so if oxygen deprivation continues, the next road from the pyruvate junction is ethanol fermentation. Ethanol is a safer option, since it's much easier for plant cells to dispose of.

Both lactic acid and ethanol fermentation pathways described above produce less usable energy than aerobic (with oxygen) respiration because they require more energy to run. For some plants, switching to this pathway during a flood means remaining alive, but growth won't resume until the water subsides.

Wait or escape?

Plants that are exposed to temporary flooding have a choice: wait it out or try to grow above the floodwater. Some save their reserves of energy and wait for the water to drop before resuming growth. Others try to get their leaves above water, where light and oxygen are present. Fascinatingly, it has recently been found out that these two strategies rely on the same hormones to produce very different outcomes.

Ethylene, a gas that is produced in plants, builds up whenever water prevents it from diffusing into the atmosphere. In the sit-and-wait plants, it stops the production of the growth-inducing hormone gibberellin. In the escapee plants, the ethylene build-up increases gibberellin, causing a growth spurt.

Eupatorium cannabinum

How is this possible? The key is in the intricacy of the network of steps that connect the hormones. In the sit-and-wait plants, one step (called submergence1A), which inhibits gibberellin, is increased by ethylene. In the escapee plants, both intermediate steps increase gibberellin.

Adaptable options

Extreme weather may turn parts of some gardens into a water frontage overnight, and dry them up at other times. Low-lying areas will always be the last to dry out, however, so they are perfect for adaptable plants that may prefer to have their feet wet. Some will survive drought using stored resources, becoming dormant before resprouting from rhizomes when the rain comes. In this chapter we will explore a range of these amphibious plants and the specialist adaptations that help them to thrive despite large variations in moisture within their environment.

Melaleuca

Micranthes pensylvanica
Swamp saxifrage

Delicate green-white flowers are held on fleshy stalks above lush green tufts of leaves. Swamp saxifrage prefers to have wet feet but can cope with short periods of both inundation and low rainfall, making it very versatile. The thickened roots store energy for fast recovery, and the upright leaf arrangement allows it to reach above shallow water.

Raising leaves into an upright position is a common strategy among plants that live next to waterways or in places where water frequently pools, for escaping short periods of flooding. This works for species that are otherwise not adapted to growing underwater (unlike those in Chapter 2). Instead of adapting their leaf types to grow underwater, these plants raise them above the water to continue photosynthesizing and exchanging gases with the atmosphere.

The leaves of swamp saxifrage are covered in hairs that can create air pockets during short periods of flooding, and also reduce water loss in drier conditions. As in many adaptable species, the thickened roots store resources to help the plant to resprout after flood damage or periods of dormancy, such as winter or drought.

This lovely plant is native to North America but has become endangered across much of its range owing to pollution and habitat loss. Providing pollen and nectar for a variety of pollinators, it deserves a place in damp gardens, where it will spread gently via offshoots and seeds.

Filipendula ulmaria
Meadowsweet

With its froth of pink or white flowers floating above vibrant green foliage, meadowsweet is a delightful plant that is native across Europe and western Asia.

Often found below the winter flood levels, *Filipendula ulmaria* thrives in wet areas.

In common with many flood-tolerant species, it has a thick rhizome, which is very good at coping with low oxygen levels in the waterlogged substrate. Meadowsweet can respire without oxygen, although it does not produce aerenchyma (air straws). Instead, it makes alcohol in the form of ethanol from stored sugars. This process creates energy, but much less than aerobic (with oxygen) respiration, simply allowing metabolism to continue so as to keep the tissues alive until the flood has passed. Creating energy in this way allows the plant to survive, but meadowsweet will resume growing only once the rhizome has access to oxygen again.

An interesting side note is the role of meadowsweet in the story of aspirin. Meadowsweet produces high levels of the hormone salicylic acid in its buds. In the 1830s salicylic acid was isolated from these buds and used to make the more stable acetylsalicylic acid. This is the active ingredient in aspirin.

In plants, salicylic acid is particularly important for responses to stress. Like humans, plants fight infection and heal through the use of antioxidants, and salicylic acid triggers antioxidant production. It also delays senescence (dying cells) in cut stems. That is why dissolving aspirin in the water of a vase of cut flowers can extend their life.

Perhaps because of their high concentration of salicylic acid, the blooms of meadowsweet are excellent cut flowers, making this plant an all-round delight for damper areas of the garden.

Lysichiton species
Western skunk cabbages

Yellow or white beacons glow from a rosette of spinach-like leaves. But skunk cabbage is not related in any way to members of the spinach or cabbage families; in fact, it is in the arum family (Araceae) and is more closely related to the popular house plants known collectively as peace lilies (*Spathiphyllum* spp.). Although flooding your house plant will no doubt kill it, skunk cabbages are happiest with a water frontage.

There are just two species of *Lysichiton*. The white-flowering one (*L. camtschatcensis*) is native to Asia, while the yellow-flowering species (*L. americanus*; shown here) is native across northwestern America. It is this yellow species and the hybrid between the two (also yellow) that are most widely grown in gardens. *L. americanus* is considered invasive across Europe (and is even banned in the UK), so, as with all plants, check the guidelines in your local area before growing it in your garden.

Although happiest in waterlogged soils, skunk cabbage is fairly unfussy about pH and nutrient conditions. But why has this lovely plant got such a stinky name? It is pollinated by a small beetle (*Pelecomalius testaceum*), which it attracts by producing an unpleasant smell. Interestingly, *L. camtschatcensis* is less smelly, as is the hybrid, no doubt contributing to its popularity in the horticultural world. However, I would still choose a boggy spot as far as possible from the house to get the most from this interesting plant.

Rodgersia species

Rodgersias

Is that a young horse chestnut growing alongside the pond? No, it's *Rodgersia aesculifolia*, so named after the similarity of its leaves to those of the horse chestnut, *Aesculus hippocastanum*.

There are only a few species of rodgersia. All originate in Asia and all are happiest in damp soil or, if the soil is dry, shaded positions, because the large surface area of their leaves allows the transpiration of high volumes of water. These plants have rhizomes – a recurring theme in this book – which allow them to store resources during dry periods.

The rhizomes that make these plants so good at growing fast and tolerating a range of conditions can also spread in the surface layers of soil. In a couple of species (such as *R. nepalensis* and *R. podophylla*), they can spread very quickly. *R. aesculifolia* and *R. pinnata* (which is common in cultivation; shown above) are more appropriate for garden settings, forming local clumps rather than spreading rampantly.

The foliage alone is a reason to grow these lovely plants, but they also produce tall stems of flowers that are held high, in shades from white to deep pink, depending on the species and variety. Rodgersia also grows well in a large pot sitting in a tray of water, so even the smallest garden can benefit from this lush and versatile plant.

Lythrum species
Loosestrifes

A sea of magenta spires lines a boggy path. *Lythrum salicaria* (purple loosestrife, shown opposite) is a stunning garden plant in the UK and is native across Europe, Asia and Australia. However, it is an ecological problem in areas of the USA, highlighting the importance of understanding locally appropriate garden plants. The good news is that there are many lovely species of *Lythrum*, each native to its own region of the northern hemisphere (such as *L. alatum*, which is native across the USA), so there is likely to be one that suits your area.

Species and varieties of loosestrife come in many shades of pink, purple and white, and most can be found in nature growing with wet feet. In wet conditions it produces aerenchyma (air straws) to carry oxygen into deeper tissues, allowing metabolism to continue despite the waterlogging.

By contrast with many completely aquatic species, however, the aerenchyma in these amphibious plants might not develop under non-waterlogged terrestrial conditions. Instead, when low oxygen is sensed, a switch turns on, causing a controlled programme of cell death in some tissues. This removes cells, creating connected spaces in the stems. It occurs most commonly in the cortex, which sits between the outer cell layer (the epidermis) and the region containing the vessels that transport water and sugars around the plant.

Taking this air-pocket process further, loosestrife produces corky tissue called phellem, which grows in a similar way to bark. This spongy substance has lots of air spaces to maximize gas storage and movement through submerged tissues. Because it has the same function as aerenchyma, this layer is called aerenchymatous phellem.

Loosestrifes also take advantage of adventitious roots (roots produced along the stems) when exposed to waterlogged conditions. These allow nutrient uptake in the more oxygenated parts of the soil or water surface, reducing the distance oxygen needs to move within the plant to reach these important organs.

Lythrum can grow quite happily in a range of soil conditions, and, as in other amphibious species, the production of aerenchyma, phellem and adventitious roots is triggered only in response to wet conditions. Under dry conditions the plant does not produce them.

This adaptability makes the various loosestrife species versatile garden plants for a changing climate, but it can also make them excellent competitors if they are introduced into an area where they are not native, so it is crucial to choose a species that is appropriate to your region. Many of the *Lythrum* species native to different regions make delightful garden plants that will pose no risk of invading the surrounding countryside.

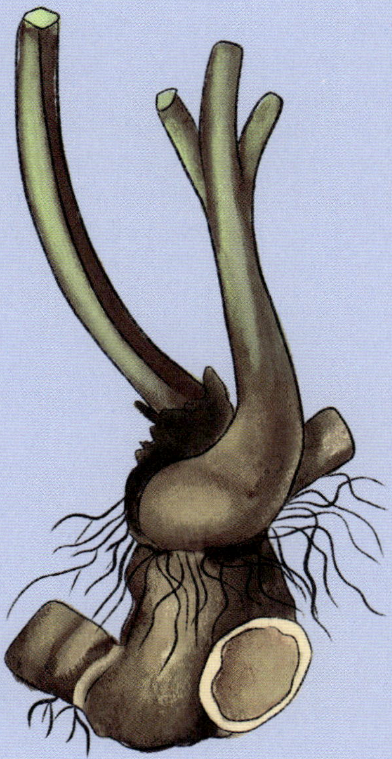

Acorus calamus
Sweet flag

Structural tufts of iris-like leaves arch at the tips. If we look more closely, yellow cone-like flowers can be seen emerging from stems, wrapped in the long leaf sheaths. This lovely plant is the sweet flag, *Acorus calamus*. The only other species in this genus is *A. gramineus*, which has thinner leaves that produce lower mounds of arching foliage. The latter, a native of Asia, has been introduced into the garden world in both glossy green and variegated varieties.

Sweet flags like to have wet feet, and will keep growing even when submerged for a time. As for many species that can grow in waterlogged conditions, getting oxygen from the air down to the roots is very important. This process begins early in seedling growth, with aerenchyma forming within the tissues. To prevent the oxygen from leaking out before it reaches the roots, an impenetrable layer of suberin develops, which prevents water and oxygen from passing through (see page 59). Oxygen moving down the aerenchyma tubes is then able to reach the roots, where it is needed for growth.

Suberin layers are very important in many environments, and they exist in almost every plant root to control the movement of water from outer root tissues into the xylem (circulatory vessels). In stressful conditions, such as waterlogging, plants can produce a second layer on the surface of the root to prevent oxygen loss.

When oxygen reaches the roots of the sweet flag, it can diffuse into the rhizosphere (the soil immediately next to the root), where it helps with nutrient uptake and the deactivation of toxic compounds present in waterlogged soils. As a result, the oxygen level directly next to the roots can be seven times higher than that elsewhere in waterlogged soil.

The rhizomes of sweet flag allow this plant to spread clonally, sending out new shoots and roots at regular intervals. Those intervals change depending on the conditions. If a patch of soil is found with the right nutrients or more oxygen, the plant can take advantage of it by sending out new growth. If the soil is less ideal, the rhizome keeps growing before popping up in a better location.

Sweet flag can also tolerate dry conditions. One way to cope with less water is to find ways to transpire less, such as producing fewer or smaller stomata (leaf pores). *A. calamus* has been shown to do exactly that, and in one case produced 100 fewer stomata per square millimetre when subjected to drought conditions. It also responds by producing fewer leaves, thus exposing less surface area for water loss, and up to a point the rhizomes may sit dormant underground until enough water is available for them to resprout.

All in all, this is an adaptable, attractive plant that will grow well in a damp area that experiences pooling water after heavy rain.

Symplocarpus foetidus
Eastern skunk cabbage

What is that stench of rotting flesh? Definitely a plant for a large garden (and far from the house), *Symplocarpus foetidus* is a fascinating, wide-ranging native of North America, Canada and Asia. Like *Lysichiton* spp. (page 64), it is a member of the aroid family, and it also likes having wet feet, although it can grow happily in merely moist conditions. It has a large rhizome for storing energy and produces aerenchyma that oxygenate the rhizome and roots.

This plant is interesting because of two adaptations in particular. The first is that it has 'contractile' roots, meaning that as the roots shrink they tighten like a spring, pulling the plant deeper into the ground. This helps to protect the stems and flower buds against cold weather and can anchor the plant in the soil, preventing winter floods or snowmelt from washing it away.

If contractile roots are not interesting enough, another amazing feature of this plant is that the flowers, which emerge early in the spring, produce heat. This heat is produced through respiration, which requires oxygen. Aerenchyma are important for this process (as well as for the respiration associated with regular development), especially if the plant is in waterlogged, low-oxygen soil, to ensure that oxygen is available for producing the heat. Often 22–27°C (12–15°F) higher than the surrounding air temperature, this heat can melt snow that might cover the flower; it also dissipates the scent to attract pollinating blowflies, including the aptly named *Calliphora vomitoria*.

Eupatorium species
Hemp agrimonies

Butterflies hover over a sea of rich green foliage topped with lacy pink flowers. *Eupatorium cannabinum* (shown here), also known as hemp agrimony, is the only species of *Eupatorium* that is native to Europe, North Africa and the Middle East; many others are found across North America (such as *E. perfoliatum*) and Asia. These plants are popular with pollinators, and many like to have wet feet.

A common theme in these chapters of damp- and water-loving plants is the presence of aerenchyma to move oxygen into waterlogged roots. *E. cannabinum* and *E. perfoliatum*, among other species, are no exception. The waterproofing layer of suberin is built into the outer cell layers, much as it is in *Acorus* spp. (page 68). This layer is thicker further from the growing root tips, ensuring that the oxygen gets all the way to the tips; there, some is allowed to leak out into the soil, creating a layer of oxygenated substrate around the growing root.

The rhizome of *E. cannabinum* – and possibly that of other, less studied species – also has this leak-proof coating, as do the lower parts of the stem, in particular when those stems are standing in surface water. All these features help to maintain internal oxygen levels to support respiration.

This resilient plant is fairly adaptable, growing well in both sun and shade and also tolerant of dry periods. The lacy flowers are a delight and attract a wide range of butterflies and other pollinators.

Alnus species
Alders

Whether forming a deep green canopy or a dense hedge, alders are adaptable trees that are well suited to swampy soil. Most species are tolerant of wet feet, but the two best adapted are *Alnus glutinosa* (a native of Europe and North Africa) and *A. japonica* (from East Asia).

One feature that helps these species to survive in waterlogged soil is the presence of lenticels – pores on the stems. To understand why these are important, let's think about the plants that we have discussed in this chapter. Most are herbaceous or low-growing perennials, with only a short distance between the leaves – where stomata can take in oxygen – and the roots in waterlogged soil. On the other hand, alders are full-sized trees in which the leaves and submerged roots are very far apart. This is where the lenticels come in. Oxygen can enter these pores along the stem, reducing the distance it needs to travel to the roots. Once inside the lenticels, oxygen travels through aerenchyma (air straws) within the stem, down to the roots.

Not only is oxygen important for normal root growth, but also in alders it feeds nitrogen-fixing microbes living in nodules on the roots. This is similar to legumes. But oxygen is needed for healthy nodule microbe communities, so submerged alder roots must ensure their snorkels are well connected.

Being also tolerant of dry conditions, these adaptable plants are valuable for gardens that are periodically exposed to extreme flooding.

Melaleuca species
Paperbarks

Brushes of stamens in a range of colours, from vibrant purple and red to white and even green, adorn green foliage atop stems wrapped in papery bark. The various species of *Melaleuca* are native to Australia, New Caledonia and the southwestern Pacific, and range from small shrubs to large trees adapted to widely varying temperatures.

Most paperbarks are well adapted to having wet feet, and some also thrive in salty water. They produce adventitious roots close to the water surface, and their tissues contain large air spaces (aerenchyma), both adaptations that we've seen elsewhere in these watery chapters, and that help to move oxygen to waterlogged roots.

But a few species, among them *M. cajuputi*, do something less expected. In common with many truly aquatic species that we covered in Chapter 2, this species – unusually for a tree – is heterophyllous. This means that when the stems and leaves are submerged, the new leaves that grow are different from the normal foliage. These underwater leaves are slimmer and have more air spaces and a thinner outer layer (cuticle) to help with gas exchange directly from the water.

Many paperbarks are also fairly tolerant of dry conditions, making them truly adaptable amphibious species. As always, check with your local authorities for risky species in your area. *M. quinquenervia* has become a serious weed across the Florida Everglades, for example, but with so many species available, there's bound to be a paperbark for you.

Rain gardens

Alive with the buzz of busy bees and flitting butterflies, a rainbow of colour stretches in front of me. Stands of drought-tolerant *Phlomis* and *Alchemilla* edge patches of *Lythrum*, *Persicaria* and *Iris*. As I cross a wooden bridge, *Acorus* stands tall between lower-growing *Lysichiton*, and *Mentha aquatica* drifts through these wetter dips.

While this beautiful rain garden has a naturalistic feel, it is in fact a deliberate and carefully designed feature, one of many by the leading garden designer Denis Dujardin. Rain gardens are crucial for slowing the movement of water through urban landscapes. After a sudden deluge, instead of all the water from rooftops and roads hitting the drains at once, rain gardens guide it through channels that are dug lower than their surroundings, feeding it into the drains gradually and helping to avoid flash flooding. Less dramatic showers are even retained entirely by rain gardens, supplying the plants with water longer into dry periods.

But what should we think about when planting a rain garden? Dujardin offers tips that apply to rain gardens everywhere. He explains that it's important to consider the position of the water table (the level of water in the soil). Beside a river or creek, the water table might be close to the surface, so moisture will never drain away. The other important factor to consider is how easily water moves through the prevailing soil. Water will drain straight through sand to the water table, for example, while clay prevents it from moving through the soil, holding puddles on the surface long after the rainfall has ended. These factors will determine which plants to choose.

In waterlogged clay soil or where the water table is close to the surface, moisture-loving bog plants thrive. Many prefer acidic soils, so using ericaceous (acidic) potting mix in the lower sections can help plants to establish. Four of Dujardin's favourites for such areas are *Iris pseudacorus*, *Acorus calamus*, *Lythrum salicaria* and *Mentha aquatica*. To provide architectural interest, tall bog myrtle (*Myrica gale*) grows happily with wet feet. Others that are happy in such conditions include *Euphorbia palustris*, *Carex nigra*, *C. elata* and *Lysichiton camtschatcensis*.

Where the water drains quickly through sandy soil, the chosen plants must be both drought-tolerant and able to cope with flushes of high water. In these conditions, Dujardin recommends trying *Myrica pensylvanica*, which is surprisingly tough, and *Carex pensylvanica*, a sedge that is native to dry forests of eastern North America.

As a parting tip, Dujardin warns that *Typha* (bulrushes) will clog rain gardens. Their leaves flop down and raise the ground level, meaning water is no longer guided through the channels designed to carry it, and instead spills over the sides.

Acantholimon

Chapter 4

Bone-Dry: Catching and Keeping Water

Introduction
Bone-Dry

Against a backdrop of deep blue skies and steep mountains, the rock layers resemble rainbows in this rainless landscape. Despite the dryness of the sandy valley floor, life abounds. Low-growing shrubs create an understorey to taller shrubs and small trees. Known as xerophytes, plants that grow in such environments have evolved some incredible adaptations to savour every drop of precious moisture, even harvesting water from the air.

Keeping cool

Places where rain is infrequent are often very sunny, and this can heat up plants' leaves and damage the cellular machinery involved in photosynthesis. As in most living things, the enzymes that are important for carrying out cellular processes have an optimal working temperature range, and they break down if they become too warm. To cool their leaves, plants often rely on water evaporating from the stomata (leaf pores). When we turn on a fan, we feel cooler because our sweat is evaporating in the moving air, using energy in the process. The same is true of leaves.

However, when water is limiting, it is far from ideal to lose moisture. To reduce the amount of light and heat in the leaf, many plants adapted to dry conditions have developed pale leaves. Some have lots of shiny white hairs, while others change the amount of coloured pigment in the leaf, including reducing the green chlorophyll that often masks the colours of other pigments. The resulting pale, silvery leaves reflect more light, which helps the plant to avoid overheating.

Ephedra

Energy sponges

Plants also produce pigments that act as energy sponges, feeding the right amount of energy to the chlorophyll for photosynthesis and mopping up any excess from around the photosynthetic proteins. Different groups of pigments often get their particular colours because they absorb light energy in one set of visible wavelengths; the remaining light is reflected, and that is the colour we see. Chlorophyll is a green pigment because it absorbs more red and blue light and reflects some of the green. Carotenoids tend to reflect orange and yellow light, which is why we perceive carrots as orange. Carotenoids (in particular those called xanthophylls) transfer some energy to chlorophyll but absorb excess energy, providing protection for the inner workings of the leaf. When there is not enough water and photosynthesis is reduced, plants often produce more of these energy sponges and may also eventually reduce the amount of chlorophyll they contain.

Eryngium

Building a boundary

When the stomata are open, one thing that determines how much water is lost is how humid the air is just outside these pores. If the air contains lots of moisture, while some water molecules will leave the leaf, some will enter through the stomata. The more humid the air, the smaller the net loss. To harness this phenomenon, many plants from dry environments create a humid layer of air close to the leaf surface, called a boundary layer. One way in which they do this is by covering the leaf in hairs, which slow the movement of air over the surface of the leaf, keeping any exiting water close to the leaf and thereby increasing the humidity in the boundary layer.

Many xerophytes (from the Greek *xeros* = dry and *phyton* = plant) also sink their stomata into the leaf, which means the epidermis is above the stomatal opening, creating a sunken pocket of humid air. Often, this pocket is surrounded by hairs or has thick ridges of waxy cuticle, all of which slow the movement of air and trap moisture near the stomate, increasing the chance that some of those molecules will re-enter the leaf.

Reducing surface area

Another way to reduce water loss from leaves is to have less leaf surface area. Many xerophytes have small or narrow leaves. Some, such as *Ephedra* (page 89) and *Casuarina* (see page 26), have taken this even further, replacing leaves altogether with photosynthetic stems. In these plants, the leaves are often tiny scales and the stems have grooves into which the stomata are sunk, presenting a very low surface area from which water can be lost.

Sports drinks for plants

As the environment dries, water becomes harder to absorb from the soil. This is because plant cells hold on to their water for as long as possible, maintaining a stable concentration of ions and salts. Meanwhile, the drier the soil is, the more ions and salts it contains, and the less water. This sets up a situation in which water is more likely to move out of the roots and into the soil by way of osmosis (see page 86). To overcome this, some plants accumulate osmolytes (just like the salts found in sports drinks) to change the osmotic balance and continue drawing water from the soil. Similar to our carefully formulated sports drinks, plants accumulate biologically safe ions, including amino acids and sugars. Called osmoprotectants, these molecules can be moved around to direct the osmotic flow of water, and many have additional functions that protect the cell from being damaged when water is low.

Digging deep

When the rain stops, the soil dries from the top down. This happens both through evaporation by wind and sun, and through uptake and transpiration by plants with shallow roots. It means that in dry soils plants must dig deeper to access water, so they often prioritize putting resources into growing deep roots. Deep-rooted trees are very important in dry climates because they can bring up water from deep reserves and release it into shallow roots – a process called hydraulic redistribution. Some of this water leaks into the soil, keeping the shallow roots alive so that they can take

Prostanthera phylicifolia

up nutrients, but also making the moisture accessible to nearby plants. This process is very widespread and is also important for replenishing deeper soil layers with water after bursts of rainfall. In that case, the shallow roots take up the water and move it to deeper roots.

Water from air

In many dry coastal areas, moist air from the ocean creates fog, or it might be cooled as it is pushed up mountains, forming low clouds. While rain may not result from these fogs or clouds, moisture condenses on leaves and stems, from where it can drip to the soil or even be taken in through leaves or aerial roots. The phenomenon is not well studied in common garden plants, but it is extremely widespread across plant families, from tall trees, such as the giant redwoods of California (*Sequoia sempervirens*), to desert mosses (*Syntrichia caninervis*) that have tiny hairs on which moisture condenses before being sucked into the plant.

The plants discussed in this chapter use some or all of these techniques to thrive in dry climates. Some also take advantage of different forms of photosynthesis (see page 28) to accumulate carbon dioxide while minimizing water loss. With increasing periods of dry weather likely in many parts of the world, there is much we can learn from these xerophytes. Used in combination with xeriscaping and sand-scaping garden designs (see page 94), these plant heroes are both tough and attractive.

Artemisia absinthium

Phlomis species
Jerusalem sages

Tiers of pink or sunshine-yellow flowers sway in the summer sun. As the seasons transition to winter, the flowers change to attractive seed heads, dusted white as though with icing sugar as the frosts come. Jerusalem sages are beautiful plants that provide structure through the seasons.

Many species of *Phlomis* are native to the Mediterranean or further east (as with *P. cashmeriana*, shown here), where hot, dry summers are followed by wet, frosty winters. To survive in these conditions, plants like those in this genus have evolved adaptations that reduce water loss and reflect extra light away.

The leaves and stems of Jerusalem sage are covered in hairs, which create a boundary layer (see page 79) close to the surface of the leaf. This means that the plant loses less water overall to the outside atmosphere, and has less need to close its stomata, allowing respiration to continue at a normal rate.

Jerusalem sages have another trick: seasonally different leaves. In the winter they grow larger leaves that are filled with water. As the conditions dry out, the plant drops these winter leaves and grows new, smaller summer leaves, with less surface area from which to lose water. These smaller leaves also contain less chlorophyll, which reduces the amount of light that can be harvested for photosynthesis. This is an advantage because it means that there is less photosynthetic machinery that could be damaged by too much light.

While photosynthesis slows, the plant keeps on hand pigment 'sponges' to mop up extra energy that could otherwise cause damage in the cells. If the weather becomes very dry, the summer leaves curl up, exposing the lighter-coloured underside to the atmosphere. The paler colour reflects more light, further reducing the chance of damage to the photosynthetic pigments.

Finally, the summer leaves can also tolerate changing shape. As water is lost from the cells, their contents shrink. Normal cells would be damaged because their contents would break away from the cell walls, but in these plants, the cell walls can flex like an elastic waistband, shrinking as water leaves the cell and swelling as it re-enters after rain.

In combination, these adaptations make *Phlomis* very tolerant of even the hottest, driest summers. Their dramatic tiers of flowers are a magnet for butterflies and other pollinators, while their structural seed heads protect insects in the colder months.

Salvia species
Sages

Lush foliage in shades of grey-green covers the sun-drenched landscape, accented here and there with pops of red or purple. Despite the plant cover, the dry ground crunches underfoot. Sages are well adapted to environments that suffer periodic drought, and, with their attractive flowers and foliage, have become widely popular garden plants.

Sages have a number of adaptations that help them to cope with drought. As the soil surface dries out, their deep roots reach water further down. Their foliage, which is often grey or silver, reflects excess light, and the hairs that cover their leaves create a humid layer of air that reduces the amount of water lost through the stomata. This reduced water loss means that the stomata can remain open for longer and photosynthesis can continue when water becomes limiting.

As the stomata close, carbon dioxide can become limiting and photosynthesis can slow down. When this happens, the incoming light energy can damage the photosynthetic machinery. To avoid this, sage leaves accumulate pigment molecules that act as sponges, mopping up excess energy. These pigments cost energy to make, so when photosynthesis is reduced, growth slows; after rain, the stomata reopen, photosynthesis increases and growth resumes.

Many sages are edible, and the flavour molecules are often enhanced when water is scarce. So, while the summer heats up and dries out, salvias bring bright flowers, lush foliage and delicious flavour.

Artemisia species
Mugworts

Silvery blue-grey feathery foliage accentuates the darker green shrubs growing nearby. *Artemisia* species are native to dry habitats around the world, and include such plants as common wormwood (*A. absinthium*, shown here), a native of North Africa and Eurasia; tree wormwood (*A. arborescens*), from the Mediterranean; and tarragon (*A. dracunculus*) and hoary mugwort (*A. stelleriana*), both from North America and Asia.

A strategy that is adopted by many plants from dry environments is to grow deep roots. This is helpful because when the rain ceases, water is depleted first in the surface layers of soil, partly by evaporation and partly from water being taken up and transpired by shallow-rooted plants. Investing resources into deep roots, as *Artemisia* does, means that water deep in the soil can be accessed during long periods of drought. While other plants are closing their stomata and shutting down growth, these deep-rooted plants can keep growing.

In *Artemisia*, thin or highly lobed leaves keep the surface area for transpiration low and create a dense, feathery appearance. Covered in fine hairs that create a moist boundary layer, the silvery leaves reflect excess light, protecting the photosynthetic machinery from damage. While these adaptations make these species well suited to dry summers, they also make them highly attractive structural plants for any garden. Many of these species also produce fragranced oils, which defend them against herbivores. Scented plants can also repel mosquitoes, and this is currently an active area of research as scientists look for natural insect repellents for human use.

Sideritis and *Stachys* species

Ironworts and lamb's ears

With their mounds of soft, woolly grey leaves, some topped with purple flowers and others creamy yellow (as in *Sideritis syriaca*, shown here), species of *Sideritis* and *Stachys* thrive in well-drained, seasonally dry conditions. These closely related genera belong to the mint family (Lamiaceae) and, like *Phlomis* and *Salvia* (pages 82 and 84), have reflective, light-coloured leaves, while their very hairy stems and leaves protect them against water loss.

When conditions become really dry, these plants have another adaptation shared by very drought-tolerant plants: they accumulate ionic molecules. Think about osmosis, a process by which water crosses a membrane to a region where there are more salts, until the concentration of water and salts is the same on both sides. Such plants as *Sideritis* and *Stachys* use this principle to direct water movement within the plant and to absorb water in the soil. But normal salts, such as sodium chloride, would damage the cells, so instead they accumulate the safer ionic amino acids and sugars. These osmolytes accumulate in the roots, pulling water from the soil. The plants can also build up osmolytes in different cell compartments or tissues, depending on where water needs to be moved to.

These adaptations to strong light and dry soil also make these plants excellent for roof gardens. Recent research has shown that they are even more efficient at cooling the substrate than the popular *Sedum* species. Versatile and attractive, *Stachys* and *Sideritis* will be successful options for a drying climate.

Prostanthera species

Australian mint bushes

A deep green, dense, small-leaved shrub with pale purple flowers fading slightly at the tips of the petals. These delicate blooms are freckled at the throat with dark purple, and on some a yellow runway guides in pollinators. *Prostanthera cuneata* is one of many members of the mint family (Lamiaceae) in this chapter that are well adapted to dry conditions. Also called the alpine mint bush, it's native to mountainous areas of southern Australia, and the leaves, when crushed, give off a delightfully minty scent.

Most *Prostanthera* species are adapted to well-drained, dry soil. Their typically small or narrow leaves reduce water loss, and are covered in microscopic balloons filled with scented oils. These tiny oil glands make the leaf surface appear bumpy, and, when held up to the light, are visible as almost clear dots.

The dense growth habit helps to keep moisture close to the plant, acting as a large boundary layer of humid air. When conditions dry further, these species close their stomata quickly to prevent water loss, and growth slows until water is available once again.

While some species of Australian mint are tolerant of some frost, others prefer mild or warm winters. *Prostanthera* species also come in reds, purples (as in *P. phylicifolia*, shown here) and pinks, all with delightfully scented evergreen leaves, making this a plant for all seasons.

Acantholimon species

Prickly thrifts

Tight green cushions cover the rocky ground. The pink flowers of *Acantholimon glumaceum* seem to pop out of beautiful white bracts that look like blooms in their own right.

Growing close to the ground as a cushion is a great way to stay out of the drying wind. To reduce water loss further, a boundary layer of humid air can be created around the stomata. One way in which *Acantholimon* does this is with lots of small hairs that slow the movement of air around them. Additionally, the stomata are sunk into the leaf, with the epidermis and thick cuticle towering above the opening. Both strategies trap moisture in the air around the stomata, preventing overall water loss from the plant.

As if dry conditions were not hard enough, many *Acantholimon* species grow in soil that contains salts. This can make it hard for the plants to take up water, which tends to move from areas with less salts to areas with more. One solution is for plants to take up salts from the soil, reducing the amount near the roots and increasing the salts in their tissues. But, as we saw with *Sideritis* and *Stachys* (page 86), too much salt is bad for the cellular functions. *Acantholimon* moves salt through the plant and secretes it from specialized leaf cells called salt glands. This keeps the salt in the cells to a safe level. It also makes this plant a delightful and useful addition to rockeries and dry gardens, including those exposed to salty sea spray.

Ephedra species
Mahuang

Striking in the contrast offered by their needle-like leaves and fleshy red seed covers, which follow the yellow flowers, *Ephedra* species are attractive structural plants adapted to dry, exposed conditions.

Many of the plants in this chapter have small or narrow leaves to keep water loss to a minimum, but *Ephedra* takes this to a whole new level by shedding its small, often scale-like leaves entirely. Instead, it has green photosynthetic stems. These thin stems are lined with furrows in which the stomata are scattered, the ridges slowing the movement of air around the stomata. To protect the stomata further from losing water, they are sunk into the stem, and the thick, waxy cuticle maximizes the boundary layer around them, preventing water loss through the surface of the stem.

Ephedra doesn't stop there. In common with many other drought-adapted plants, it accumulates osmolytes, which draw water into cell components where it's most needed. When conditions dry up, it builds up an osmolyte called proline to a much greater extent than when water is plentiful.

Another feature of this interesting plant is that it avoids putting its roots into the top layers of soil, instead sending them deep into the substrate. These deeper layers of soil hold on to water for longer, and deep-rooted plants can access that water, continuing to grow as the surface dries out.

The pollen of these fascinating gymnosperms (plants with exposed seeds) is held along the stems like mini petal-less flowers, and the fleshy red seed cases look like berries, making this plant both attractive and tough enough for long, dry summers.

Euphorbia species
Spurges

The sun beats down on a rocky landscape. Shrubs with succulent, angular stems make interesting shapes in this harsh environment. Believe it or not, some of these cactus-like succulents belong to the same genus as poinsettia and the popular garden spurges. As one of the largest genera of flowering plants, it's perhaps not surprising that species of *Euphorbia* have evolved to survive in almost every environmental condition in the world.

Succulence – the ability to store water in the tissues – appears in a range of ways in *Euphorbia*, even in the leafy garden spurges. Some store water in stems that appear almost 'normal', or in slightly thickened leaves, while others have the obviously fleshy leaves and stems typical of the most drought-tolerant succulents.

The genus *Euphorbia* is also the only one to make use of all three types of photosynthesis (see pages 20–1). As a quick reminder, C3 is the most common type and involves fixing carbon dioxide via the enzyme RuBisCO to make molecules that contain three carbons (hence C3). For C3 photosynthesis to occur, the stomata must be open, so many of these plants will lose more than 95 per cent of the water they take up. To overcome this, some plants concentrate the carbon dioxide around RuBisCO, meaning the stomata can close sooner. C4 plants do this by fixing carbon dioxide into a molecule with four carbons before moving it to specialized cells containing RuBisCO. Many spurges use this method, including two of the very few trees that photosynthesize using the C4 process (*E. olowaluana* and *E. herbstii*, both of which are from Hawaii).

CAM photosynthesis is even better at reducing water loss because it allows the plants to open their stomata only at night. During these cooler hours, carbon dioxide can enter the stomata and is fixed into four-carbon malic acid, which can be stored until the daylight hours. The stomata close as the sun rises, and carbon dioxide is released from the malic acid and fed to the RuBisCO. While this method is extremely good at reducing water loss, it costs more energy and there is a limit to how much carbon dioxide can be stored. This means that CAM plants grow more slowly.

Many species can switch between C3 and CAM, depending on the conditions. When water is plentiful, they use C3 photosynthesis to grow more quickly, and when the soil dries out, they change to CAM, reducing their water use to survive even the driest conditions. Many of the succulent *Euphorbia* species make this switch between CAM and C3 photosynthesis, while the desert species use only CAM.

Whether your garden is desert-dry or just suffers occasional summer droughts, or indeed for just about any other conditions, there's a spurge ready to grow. With what look like bright flowers in oranges, reds, yellows and pale greens (but are actually modified leaves), these fascinating plants come in a variety of structurally interesting shapes and forms.

Grevillea species

Spider flowers

Spidery flowers adorn deep green feathery foliage. Almost all species of *Grevillea* are native to Australia, adapted to a range of temperatures and typically preferring dry, well-drained soil. The variety of colour is extensive and the shapes include bunches of frothy flowers on *G. crithmifolia*, sprays of spidery flowers on *G. rosmarinifolia* (shown here), and the longer flower heads of such species as *G. pteridifolia*.

Deep-rooted and with narrow leaves to minimize the water lost through transpiration, these plants can keep growing for longer than many as the weather dries up. Well-drained soil is often fairly low in nutrients, so to overcome this, spider flowers produce special structures called cluster roots to help them take up essential minerals, in particular phosphorus. When the shoot signals that phosphorus is low, the roots develop brush-like bunches of small roots, the higher surface area making the most of any phosphorous in the soil. But phosphorus is often bound to the soil and hard to take up, so these cluster roots exude citric acid, which releases the phosphorus from the soil and makes it available to the plant.

Both tough and beautiful, these evergreen bushes have flowers that are magnets for birds and other pollinators. There are many hybrids and varieties, so there's bound to be a shrub of the right size with a bloom for any colour palette and any garden.

Eryngium species
Sea hollies

Purple flower heads are framed by stunning spiky, vampire-style collars. Sea holly species are native to many climates around the world, some found in rocky or coastal regions and others in grassland. Their fascinating collars, which range from wide and spiky to fine and feathery, and from purple and green to silvery white, have led to their widespread use in gardens.

One feature that makes species in this genus well adapted to dry (and in some cases cold) climates is that they are hemicryptophytes – they have the ability to remain dormant just beneath the soil surface. (A full cryptophyte lies dormant deeper in the soil.) Maintaining a series of dormant buds just below the surface protects them from drying out or icing up, and means they can resprout quickly, outcompeting slower-growing plants.

Many species of *Eryngium* also have deep, thick roots that can both store energy and access deeper water sources. *E. maritimum*, for example, is native to rocky coastal areas of Europe and the Mediterranean, and has roots that can reach as far down as 3m (10ft) below the surface.

In common with other drought-adapted plants, many types of sea holly have light-coloured leaves to reflect excess sunshine and protect the photosynthetic machinery. The leaves are also covered with a thick waxy layer to reduce water loss from the leaf surface.

Decorative and structural, the flowers and seed heads remain on their collars into the winter months. This is truly a plant to enjoy all year round.

Xeriscaping

Boulders softened by desert ferns, Pacific Coast irises and *Phlomis* create miniature mountain landscapes. Across the path, another rocky landscape sports spiralling succulents, while spiny yuccas stand guard, their radiating leaves creating beautiful spheres.

These two landscapes at Cistus Design Nursery in Portland, Oregon, are gardens designed for xeriscaping (dry landscapes). But while both contain drought-tolerant plants, they need rain during different seasons. The boulder garden is happy with months of rainless days in the summer, while the succulent garden prefers hot summers accented by sudden storms, so it needs watering every few weeks.

Sean Hogan, founder and principal designer of Cistus, explains that every plant has a set of conditions that it can tolerate, and others that it is really happy in. Without summer watering, for example, yuccas will die in the dry summers of the US Pacific Coast, which is classified as a Mediterranean climate. It's all about knowing the local rainfall patterns, he explains, choosing plants that are best suited to those conditions, or accepting that some watering may be needed at particular times of the year.

Still, xeriscape gardens, unlike lawns, are happy to wait for weeks between drinks, rather than needing to be watered every few days. Hogan explains that it's hard for people to stop nurturing when that has been 'normal' garden practice, but with frequent watering many natives or Mediterranean plants will die, and those that are not dead will look as though they wish they were. These plants are adapted to dry summers and have no adaptations to cope with frequent summer rainfall.

The other key aspect of choosing plants for dry gardens is to know your soil type, then check what kind of soil a potential plant grows in in the wild. Knowing whether it is happiest in sand, clay or rock is an excellent start to getting the right plant in the right place, and much easier now that so much information is available online.

The fun part is choosing the plants that will not just survive but thrive. And gardens should be fun, Hogan reminds us. Native plants don't have to be a punishment; they can be planted in blocks of colour and with interesting textural combinations. Local plants can be exciting, and in the northern hemisphere there are evergreen versions of just about anything.

As in rain gardens (page 74), a dry garden can have different areas where plants that prefer similar conditions grow together. In climates with dry summers, for example, yuccas and agaves can be planted together and watered every couple of weeks, while plants adapted to Mediterranean climates can be left to dry out over the summer. Some of Hogan's favourite drought-tolerant plants for sporadic summer rain are *Yucca rostrata* and *Agave ovatifolia*. Other beautiful examples are the desert ferns, such as *Hemionitis lindheimeri*, which is orange underneath and silvery on top and, when planted in a hanging pot or on a slope, gives a stunning bicolour effect.

Plants that are adapted to long summers include any manzanita (*Arctostaphylos*), with their stunning barks and long flowering seasons, and *Ceanothus*, the California lilac. Planting swathes of *Phlomis* and rosemary is not just attractive but also minimizes the space for weeds. Such species as *Frangula californica* (coffeeberry) and *F. c.* subsp. *tomentella* have small flowers that are magnets for pollinators, and berries that birds love later in the season. Even the driest garden can be filled with colour and texture if we choose 'plants for the place' and play with their groupings to make interesting, lovely and, above all, fun designs.

Juniperus communis

Chapter 5

Ice-Cold: Winter Survivors and Thrivers

Introduction
Ice-Cold

A thick layer of snow covers a frozen landscape. Although the temperature is now well below freezing, just a week ago this was a warm, late autumn scene. Along with extremes of heat, flood and drought, sudden dips in temperature are increasingly likely in a changing climate. Some plants are better able to withstand them than others, but first we must understand the challenges freezing weather brings.

Ice, ice, baby

As the temperature drops below 0°C (32°F), water freezes, changing from a liquid to a solid. Plants' roots are able to take up water only when it is liquid, so when ice forms, that moisture is no longer accessible to them. This means that while there may be ice all around or snow on the ground, the conditions are effectively the same as those of drought.

Ice causes other problems, too, such as bursting cells and damaging important membranes. Think of a plastic bottle of water in the freezer with its cap on; as the ice forms, the bottle swells until it bursts. This is because ice molecules are less closely packed than the molecules of liquid water. It also means that ice is less dense than water, which is why it floats. In plant tissues, the formation of ice increases the volume of the cells, bursting any membranes that are too rigid, while large ice crystals can puncture the cells.

To overcome these challenges, plants have found ways to prevent ice from forming by making antifreeze and finding other ways turn up the heating.

Draba

Antifreeze

While pure water freezes at 0°C, that freezing point is lowered if other molecules are mixed with it. Saltwater, such as that in the sea, freezes at about -1.8°C (28.7°F), and this is why roads and pavements are salted in the winter to prevent the formation of ice. Another example is ethylene glycol, which is added to coolant in cars as an antifreeze. Ethylene glycol is an alcohol, and alcohol has a much lower freezing point than water. When it is added to the coolant water, the freezing point of the water drops to -45°C (-49°F).

Plants take advantage of this property of water by accumulating salts, sugars and alcohols in their cells – effectively making their own antifreeze. While some of these molecules prevent ice from forming in the first place, others attach to minuscule ice crystals and prevent them from growing into larger crystals that could puncture the cells.

But ice is not the only problem in low temperatures. As cell membranes approach freezing, they become less flexible. When ice forms and the cell contents swell, the change in volume can crack holes in the membranes. These holes are a problem because the important cell contents can leak out of them and parts of the cell that should be separate can mix, killing the enzymes that make all the growth processes happen. Plants that are adapted to cold overcome this by accumulating fatty molecules called sphingolipids in their membranes to keep them flexible.

Some flowers and leaves are able to continue their functions even after they have been completely frozen. They do this by combining these strategies: preventing large ice crystals from forming in favour of tiny, slushy crystals using antifreeze-type molecules, and using fatty moisturizers to keep the cell membranes flexible. In this way the cells can ride out the swelling caused by freezing, keeping their enzymes and other active components safe.

Helleborus orientalis

Central heating

The processes associated with metabolism and growth in all living things create heat energy. Our bodies generate energy to maintain our core temperature at 37°C (98.6°F), increasing metabolic processes to keep our temperature stable when the air outside our body is cold. While plants don't maintain a stable core temperature as humans do, they do create heat as a by-product of the metabolic processes they use to grow.

Observed through a thermal camera on a freezing night, plants will appear warmer than the background. They can use this to help them in colder conditions. One strategy is to grow very dense above-ground stems and leaves. These hold on to the heat generated through metabolism and respiration for much longer, as the heat takes longer to dissipate. This central heat can also be insulated underground in large rhizomes, bulbs or tubers. The ground acts as a blanket, especially where there is a thick layer of leaf litter, cooling much more slowly than the air as the winter sets in (or during a cold snap in other seasons). Tucked below ground, the large volume of storage organs, such as bulbs and rhizomes, means they lose heat more slowly than thin roots do, and they are filled with the antifreeze sugars that reduce the chance of ice forming in the cells. While the above-ground leaves and stems may die back, the rhizomes remain safely resting, ready to sprout new leaves and stems when the warmth returns.

Microbes and fungi also create heat through metabolic processes, and recently it has been found that plants can exploit this. If we take into account the tiny size of microbes, they produce far more heat than larger organisms – and just how much depends on their food source. High-energy sugars, such as sucrose, are produced in the nectaries of many flowers, and yeasts (single-celled fungi) thrive on this food, growing and multiplying – and producing heat. Winter-flowering plants, such as hellebores (page 104), exploit these tiny heat-producing organisms to prevent their flowers from freezing, and reward their pollinators with a warm stopover.

Eranthis hyemalis

Avoiding the big freeze

One of the most obvious aspects of the turn of season from summer to winter, particularly in the northern hemisphere, is the change in leaf colour before the leaves of deciduous plants fall. These colours are caused by light reflecting off pigments that are present in the leaves throughout the year, but are drowned out by the more dominant green of chlorophyll. Chlorophyll contains quite a lot of nitrogen, which is taken up from the soil and can be a limiting resource. So, as the weather cools, the plants take this nitrogen back into the stem and roots, storing them until the spring. This reveals the other pigments, producing the beautiful colours of autumn.

Why drop leaves at all?

There are many reasons why some plants have evolved to be deciduous. On the one hand, it lessens wind resistance, minimizing the chance of a tree being blown over in winter storms. Another advantage is that it reduces water loss. Having large leaves is great during the long summer days, allowing the plant to capture large amounts of light and grow fast. Dropping those big leaves prevents water loss during the months when water can be locked up in ice. The fallen leaves also help to insulate the ground and protect overwintering organisms – so, when the cold months approach, let them be!

Ilex aquifolium

In this chapter you will find plants that bring bright colours to autumn, winter and very early spring through flowers, berries, evergreen shapes and textures, and glowing leaves. But not forgotten are such summer joys as hardy geraniums, with their underground rhizomes that keep them safe over the winter.

Cyclamen hederifolium
Cyclamen

Like candle flames, the flowers of cyclamen shimmer pink and white above a carpet of dark green and variegated foliage. Late autumn-flowering *Cyclamen hederifolium* has delightful silver-green variegated leaves, reminiscent of ivy (*Hedera helix*; page 132) – hence its name. *C. coum* flowers in the winter above rounded leaves, some patterned with lighter green.

As geophytes native to the Mediterranean, cyclamen have large below-ground corms that store energy during the dry summer, when the leaves have died back. In the autumn, new shoots sprout from the bulb in preparation for the autumn–winter growing season. Of course, having an underground storage organ is not only useful for waiting out dry periods; it also protects below-ground reserves of energy against extremes of cold. The larger the corm, the longer it will stay warm, and this is especially true of those that are actively metabolizing and growing during the cold months.

In common with many drought-adapted plants, cyclamens accumulate soluble sugars and osmolytes (see page 80) during their growing season to maximize water uptake and reduce damage to their membranes. These sugars and osmolytes also reduce the temperature at which water freezes in the leaves and flowers.

Advertising flowers above the foliage helps pollinators to find them. After pollination, the petals drop off and the flower stalks curl into tight springs, bringing the developing seed closer to the plant and sheltering it from cold. When ripe, the seeds are coated in a sugary substance that attracts ants, which take the seeds away and kindly plant them far from the parent plant. So don't be surprised if a new pink flame pops up to bring winter colour to a distant patch in the garden.

Viola species
Pansies/violets

Available in a rainbow of colours, pansies are a popular winter-flowering bedding plant, with cheery faces that brighten the dreariest day. While pansies are large-flowered annuals or biennials, the genus *Viola* also includes the smaller-flowered, often perennial violets. Despite their name, violets come in shades of white, yellow, pink and, of course, purple.

Species of *Viola* are found all over the world in many habitats, including alpine bogs (*V. palustris*) and cold deserts (*V. atropurpurea*). To prevent their cells from freezing, they accumulate molecules, such as amino acids and sugars, that act as antifreeze and protect their membranes. These protective molecules are similar to those found in desert plants. Frozen climates can cause drought stress in plants because the roots are unable to take up water from frozen soil, and the membranes can become strained as the cell contents shrink or swell. The protective molecules, while lowering the freezing point of water to prevent icing, also direct the movement of water in the tissues by influencing osmosis, protecting the membranes from damage as the cells shrink through water loss.

Beautiful yet tough, pansies and violets – such as the sweet violet, *V. odorata*, shown here – also have edible flowers that are ideal for decorating salads and cakes. Plant these species in a pot outside the kitchen, ready to bring to the table even in the depths of winter.

Helleborus species
Hellebores

Discs of coloured flowers stand among mottled dark green foliage, bringing sunshine to the short days of winter. Hellebores come in an ever-increasing range of colours, from whites (as in the *Helleborus orientalis* shown here) and greens to bright pinks, reds and purples, and even the black of night.

These useful plants are happy in shaded positions, and the foliage is also attractive, in lush mottled greens. Hellebores grow below-ground rhizomes in which they store energy, insulated by the soil to protect against extreme changes in temperature.

The lovely flowers, which appear in the winter and early spring, are pollinated by bumblebees. Blooming during the cold months means that there is less competition for pollinators, increasing the chance that pollen will be transferred directly between two hellebore plants and thus helping with seed production.

Being winter flowerers, hellebores need strategies to prevent ice from forming in their tissues and to maintain a temperature that is suitable for pollination. In common with many plants, they produce sugars and other molecules that act as antifreeze. But hellebores have another trick, too. They support a dense community of yeasts, which helps to warm the flowers.

Yeasts are single-celled fungi that come in many different species. Baker's yeast (*Saccharomyces cerevisiae*) is a familiar example that feeds on sugar in bread dough, producing carbon dioxide, which makes the dough light and airy. The process yeast uses to make carbon dioxide is a form of respiration, similar to that found in waterlogged plants (see Chapter 3), and it also creates heat. The faster the growth of the yeast, the higher the respiration and the more heat it produces. The yeast in the nectaries of hellebores (in this case *Metschnikowia reukaufii*) feeds on the sucrose-heavy nectar, which supports the yeast's fast growth and reproduction. This in turn heats up the nectary and surrounding air.

Increasing the temperature of the flower has many potential benefits. It supports the growth of the pollen tube and successful fertilization, which can fail if temperatures are too low. It may also help pollinators find the flowers through air currents (convection) created by the warmth. Additionally, when pollinators visit these flowers, they are rewarded with a pulse of warm air that boosts their body temperature. Taking advantage of this, some bumblebees will choose the warmest flowers, keeping most of their body within the airspace of the flower to maximize the energy they can absorb.

Hellebores come in a range of hybrids, including a few with double flowers, some that hold their blooms above the foliage, and others that produce flowers among the leaves or even, in deciduous varieties, straight from the ground. Their bright, cheerful blooms that bring warmth to bees can bring an illusion of summer to any garden.

Eranthis hyemalis
Winter aconite

Even during the short days of winter, patches of bright yellow light up deciduous woodland and gardens. One of the earliest plants to flower, this little aconite is a wonderful addition to a winter garden.

The cold-tolerant flowers emerge through melting snow, each surrounded by a collar of leafy green bracts. Their generous nectaries reward pollinators in these cold conditions.

Winter aconites are winter ephemerals. This means that they grow quickly in order to take advantage of sunlight streaming through the leafless branches of deciduous trees and the almost non-existent competition for pollinators' attention. Their leaves soak up the sun and send sugars to tubers below ground, where they are stored as starches throughout the summer and autumn. This energy store is then ready for the supercharged growth season typical of ephemeral plants.

Although a native of the Mediterranean, *Eranthis hyemalis* has become naturalized across northern Europe, the UK and parts of North America, and is popular in horticulture. It was introduced into the UK in the late 16th century and had gained popularity by the 18th century, when naturalistic plantings were the fashion. It was also introduced into the US at this time. No doubt part of the plant's popularity is its resistance to deer and squirrels, thanks to toxic compounds that deter herbivores.

Being members of the buttercup family (Ranunculaceae), many species of *Eranthis* have buttercup-yellow flowers, but there are a few with white blooms, such as *E. pinnatifida* and *E. albiflora*. All bring a welcome splash of light when there are few other flowers about.

Scilla siberica
Wood squill

Drifts of blue appear under the leafless trees in early spring. *Scilla siberica*, in common with other species of *Scilla*, has below-ground bulbs that store energy during dry summers and cold winters, supporting fast leaf growth in the late winter followed quickly by carpets of blue or white flowers.

Most plants try to avoid flowering in the winter to protect their important reproductive organs from freezing, but early blooming does have its advantages. The leaves can soak up the sun before the trees' leaves grow and the canopy closes overhead, and there is less competition for pollinators.

Since it grows and flowers in late winter and early spring, the wood squill is an example of a plant that must prevent or survive freezing. As discussed on page 99, when ice crystals form in tissues, the cell membranes burst, killing the tissue. One strategy the wood squill makes use of is to accumulate sugars and other proteins to lower the temperature at which ice forms.

Reproduction also suffers in cold conditions. Pollination relies on a pollen tube growing from the pollen grain into the female ovule. In most plants, when the temperature becomes too low (approaching 0°C/32°F), the pollen tube is unable to grow. But in wood squill, pollen tube formation continues well below freezing. Even more remarkably, the flowers can withstand temperatures below -20°C (-4°F) and subsequently resume growing.

Recent research suggests that microbes help the plants to cope in such conditions. Microbes collected from *Scilla* plants and given to green beans improved the beans' tolerance to freezing. Exactly how this works in *S. siberica* is still unknown, but plant–microbe interaction is an exciting and active field of research.

Draba species
Whitlowgrasses

With their small, densely packed leaves, whitlowgrasses cling to the ground, creating a green carpet that softens the rocky landscape. Many species of whitlowgrass are found in alpine rocky environments and have developed low-growing habits to protect themselves from the full force of icy winds. The dense growth form also helps to keep any warmth created by metabolism within the tight ball of leaves. These mountainous habitats with their shallow pockets of soil often experience dry periods, and the small leaves and dense growth also help to reduce surface area for evaporation.

One species, *Draba chionophila*, is found far up in the Venezuelan Andes, the highest-growing vascular plant (with fluid-carrying vessels) in this region. At more than 4.5km (almost 3 miles) above sea level, this plant must be able to survive in extremely cold conditions. With its cylindrical, semi-succulent leaves, it can accumulate high levels of sugars and amino acids that guard against excessive osmosis and help it to survive dry periods. These compounds protect the membranes from damage as the cells swell and shrink, meaning that the plant can freeze and thaw, resuming growth when conditions become favourable once more.

The cushion-like growth habit makes these plants ideal for rocky gardens of any size. The dense leaves of the different species encompass flat rosettes, tiny, round semi-succulent leaves and spiky-looking tufts – all attractive in their own right. Topping the foliage are charming yellow (as in *D. aizoides*, shown here) or white flowers, some close to the leafy cushions, others waving above.

Geranium species
Hardy geraniums

A splash of colour appears in the summer on a backdrop of divided leaves that grow as a dense mound. Hardy geraniums, also called cranesbills, can cope with cold winters and are also surprisingly drought-tolerant.

It's important to start by clarifying two groups of plants that are often confused: geraniums and pelargoniums. Pelargoniums – popular bedding plants – are often called zonal geraniums or just geraniums, but in fact they belong to a different genus: *Pelargonium*. Hardy geraniums, on the other hand, are in the genus *Geranium*.

The two groups have similarities, since they are both in the family Geraniaceae, but there are also differences. Geraniums typically have five petals of equal shape and size, forming a symmetrical flower. Pelargonium flowers are often asymmetrical, with different shapes or markings on some petals and not on others. Pelargonium plants are also often thick and fleshy – sometimes called semi-succulent – which geraniums are not. Pelargoniums, although tolerant of dry summers, cannot cope with cold and must be protected in the winter. By contrast, geraniums are cold-tolerant and can be left all winter to rest in their garden bed.

Hardy geranium species drop their leaves in the very coldest weather. Their rhizomes store energy below ground, protected from swings in temperature and ready to sprout new leaves once the warmth returns. It is these rhizomes that make hardy geraniums very tolerant of dry weather.

Colour options for true geraniums range from pastels to bright pinks and purples, some with veins of colour on white or pastel petals, and the flowers can be small and delicate (as in *G. maculatum*, shown here) or large and bold. With varieties that are happy in sun or shade, these delightful and adaptable species are valuable garden plants.

Crataegus species
Hawthorns

Clouds of magenta cascade over the fence, carpeting the footpath with pink petals. *Crataegus laevigata* 'Paul's Scarlet', one of several pink varieties of hawthorn, is a delightful small tree for urban gardens. Many other hawthorns have white blossoms, dusting the hedgerows with 'snow' in the spring, while others are variable (as *C. oxyacantha*, shown here). There are many species that are native across the northern hemisphere, including red hawthorn (*C. mollis*) and Washington hawthorn (*C. phaenopyrum*).

Hawthorn is very tolerant of cold winters, and accumulates protective compounds to avoid damage to the cell membranes, and antioxidants to mop up high-energy molecules that might otherwise damage cell processes. More of these high-energy molecules are produced when the plant is under stress, so when they are stressed, plants also produce more antioxidants to mop them up. This is the same in humans, and the reason we might take the antioxidant vitamin C when we have a cold. The fruit – called haws – of some hawthorn species, such as *C. monogyna*, are edible, and the protective antioxidants they contain can also benefit people. The haws are also high in pectin, making them perfect for jams and jellies.

Hawthorns are valuable for attracting wildlife into gardens. The masses of flowers bring pollinating insects, and the berries delight birds in autumn. Whether loosely pruned in natural-looking thickets or trimmed to dense hedges, the foliage is a haven for nesting birds.

***Juniperus* species**
Junipers

Blue-green foliage is emphasized by blue berries that cluster along the stems. Species of *Juniperus* are found across the northern hemisphere, some adapted to the tropics and others growing inside the Arctic Circle.

Junipers are conifers, which means they do not produce flowers or fruit. Juniper 'berries', then, are actually seed cones, the scales that make up the cone being fleshy and fused to look like a berry. Have a close look at them next time you come across a juniper bush to see the creases where the scales fuse. The unripe (green) berries are used to flavour gin, and the ripe berries to season meat dishes.

To grow juniper for its berries, check whether the species is monoecious (produces male and female flowers on the same plant) or dioecious (male and female flowers on different plants). *J. communis*, the common juniper (shown here), is dioecious, so female plants are needed to produce berries, and a male must be growing nearby so that the wind can blow the pollen to the female.

Similar to holly (page 112), many junipers produce two leaf types (heterophylly): spiky juvenile needles and softer, smaller scales on older branches. Protective waxes on the leaves prevent water loss during dry summers or during the winter, when frozen water is unobtainable by the roots.

In the garden, these tough, attractive, structural plants come in all shapes and sizes, from large trees to small shrubs, and upright or prostrate (ground-hugging). Their evergreen foliage provides winter interest and shelters overwintering wildlife.

Ilex aquifolium
Common holly

With its glossy evergreen foliage and bright red berries, common holly is a staple of winter wreaths and Christmas carols. Its undulating, spiky leaves are dark green or variegated, and some varieties are even tinged with shades of pink or purple.

But that's not all. Large, mature holly plants also have leaves with smooth margins and no sharp teeth, usually on the upper branches. This is called heterophylly, from the Greek *heteros* (different) and *phyllon* (leaf). Many amphibious plants (see Chapter 3) also display heterophylly, with one type of leaf underwater and a different type above the surface. But holly isn't amphibious; in this case, the sharp, tough leaves are a way of discouraging herbivores from feeding on the young trees.

Holly's leaves are also able to resist freezing by accumulating protective and antifreeze molecules. As the winter sets in, the leaves accumulate more and more of these compounds, increasing their freeze tolerance from -9°C (15°F) at the beginning of autumn to well below -24°C (-11°F) in midwinter. These sugars are so good at protecting the membranes during freeze–thaw changes that photosynthesis can continue throughout the winter, and holly leaves tend to remain on the tree for more than five years. These glossy leaves filled with protective molecules also help holly to tolerate drought, making it an adaptable option for many gardens.

Another interesting thing about holly is the process of seed germination. Holly produces a berry that is a favourite of birds in the winter months. For berry-producing plants to take advantage of birds for seed dispersal, the seeds must be able to survive passing through the bird's digestive system, and to this end holly seeds have a thick coating that is broken down slowly as it travels through the bird's intestine until it emerges once more, ready to germinate. Without the bird's digestion to strip the seed of this coating, the gardener must scrub it off the seed, a process that is called scarification.

But this is just the beginning of growing holly from seed. The next step is to wake the seed from dormancy. Many seeds rest until the temperature or moisture conditions are just right for germination, and for many plants, the seeds must go through a period of cold (called stratification) before the return of warmth prompts them to germinate. For holly to break dormancy and germinate, the seeds require not one winter, not two, but three!

Great whether trimmed as a hedge or left as a small tree, holly is an interesting and tough species that is suitable for many locations. Its evergreen leaves and bright red berries bring structure and colour in the winter, and it is great for wildlife, feeding birds when food is scarce and providing dense, protective foliage for overwintering wildlife and spring nesting. Choose a variegated variety to brighten up a dark corner of the garden, but don't forget the gardening gloves when preparing those Christmas wreaths.

Gardening in a winter wonderland

Like the flames of a campfire, stems of yellow, orange and red flicker in the deciduous planting, accented with the white stems of silver birch, which will soon reflect the colours of the sunset. But not all is deciduous or lying in slumber. Winter-flowering heathers paint the landscape in every shade of pink and white.

Despite the dormancy of the colder months, gardens can still be filled with interest and colour. Even in places where snow covers the ground, shrubs and trees with unusual shapes and colourful bark create a winter wonderland. In areas where frost glistens and everything appears bedecked with tiny diamonds, last year's seed heads become royal sceptres or magic wands shimmering in the soft light. Autumn leaves scattered across the ground, long drained of their colour, now sparkle with ice crystals and crunch deliciously underfoot.

To bring life and interest to the winter months, it helps to think about structure, texture, colour and scent. Structure is the bones underneath everything else, and texture creates patterns on or around those bones. Most of us associate colour and scent with spring and summer gardens, but they can be the baubles on the Christmas tree in winter.

Trees provide a garden with good bones, but for this purpose they don't have to be big. Many lesser trees offer lots of interest, and their small size allows them to become a centrepiece. Some, such as the corkscrew hazel (*Corylus avellana* 'Contorta'), have branches in interesting shapes. Between the trees or in more open sections of the garden, adding fences, piles or rock walls – anything that adds underlying structure – will create an interesting landscape that is more on show in the winter. Finally, instead of cutting back plants after the first frosts, leave old seed heads and the 'dead' stems of annuals or deciduous perennials. Ice crystals can form and shimmer on these structures, while any remaining seeds provide valuable food for birds, and insects can shelter in the small hollows that form as the stems dry out. Unusual logs and stones can bring structural interest to winter gardens, as can evergreen shrubs and climbers growing on walls and through trees.

Intriguing seed heads, such as the almost transparent seed cases of honesty (*Lunaria annua*) and the chandeliers of cow parsleys (*Anthriscus* spp.) and *Allium*, add texture and pattern. Bark can also bring great interest; that of cherry trees (*Prunus* spp.), for example, is shiny and copper-coloured. Birches can have shiny coppery bark too, or white, as with *Betula pendula* and *B. utilis*, while snow gums (*Eucalyptus coccifera*) have smooth, rainbow-coloured bark. Manzanitas (*Arctostaphylos* spp.) have beautifully smooth orange to red bark that peels into tight curls, revealing the rich colours of the even smoother layer underneath. Papery bark adds a different texture; that of the paperbark maple (*Acer griseum*), for instance, is flaky and cinnamon-coloured, while the Peking lilac (*Syringa reticulata* subsp. *pekinensis* 'Morton') has almost *Prunus*-like bark that peels away.

In talking about the texture of bark, it becomes clear that it can also bring colour to the garden. Adding to the list above is *Cornus sanguinea*, with its stems in flames of yellow, orange and fire-engine red. There are also a surprising number of winter-flowering plants, such as cyclamens (page 102) and anemones, that disappear during the warmer months. Other winter flowerers include hellebores (page 104) and many heathers. Bright pops of colour can also come from berries, including the red of holly, such as *Ilex aquifolium* (page 112) or *I. verticillata*, *Viburnum opulus*, many species of *Sorbus* (the rowan or mountain ash), the orange-red of *Pyracantha* and *Cotoneaster*, the pink of *Symphoricarpos orbiculatus*, the pink and blue of *Viburnum nudum*, and the purples of *Callicarpa* spp. Whatever your colour interest, there's a berry to match.

Finally, the sweet, clear scents of *Daphne* and *Sarcococca* fill the crisp winter air. Completing the sensory experience, overwintering birds chirp from berry-laden trees, bees buzz around the winter flowers, and frosty leaves crunch underfoot. There is a great deal to love about a winter garden.

Brachyscome iberidifolia

Chapter 6

Limited Living: Adaptations for Shallow Soils

Introduction
Limited Living

As I walk between tall stone walls, ferns reach out from the lower cracks, while flowers higher up spill into the sun, advertising their nectar to passing pollinators. Weaving between them, delicate creeping plants grow in seemingly impossible positions. At the end of this green pathway, the sun streams on to a rocky garden filled to bursting with glorious colour and fascinating foliage. Lizards warm themselves on areas of bare rock, while butterflies stretch their wings in the sun atop last year's seed heads.

These cracks and crevices are valuable refuges for wildlife, and the sunny vistas are perfect for cold-blooded organisms to warm their bodies, but the plants that grow in such places need to be surprisingly tough.

Craspedia

Microsites: crevice climates

While we might see a range of plants growing in a large rockery or up a high wall of rock or stone, each individual plant experiences different conditions. In an apparently sunny rockery, for example, the uneven structure means that there will be spots that are shady in the morning but bake in the sun later on, or that might be in the morning sun but shaded during the main heat of the day. A shower of rain will become a torrent of water funnelled through the cracks in the rocky surface, tugging at the plants and threatening to dislodge them. Almost as quickly, after the shower passes, the heat rises and the life-giving water evaporates. One spot could be exposed to the full brunt of the wind, while another might be protected by the rocks around it. These varying spots are called microsites, and in addition to the hazards just mentioned, the amount of sun or wind they receive can change dramatically across the year.

Another common type of rockery occurs in shaded areas near waterways, or in that corner of the garden where the tap or water tank stands. These areas might never see direct sun, but still the cracks and crevices can be filled with plant life. Ferns and mosses, in particular, love these shady, damp and rocky positions.

For the more exposed positions in a rockery, plants commonly referred to as alpines are often ideal. They evolved in rocky mountain areas and like well-drained sandy or limited soils. But plants don't have to be alpines to love rockeries and crevice gardens; many coastal plants adapted to headlands or sandy soil also work well, as do many desert plants.

Adaptations to prevent cracking up

The plants that are best adapted to survive the swings in conditions most commonly associated with well-draining soil are able to survive periods between rains. We've already seen many of these adaptations when exploring drought-tolerant plants (see Chapter 4): the thick, waxy cuticle on the leaves to prevent water loss; succulent leaves to store water; silvery foliage that reflects extra heat; and hairy leaves that create a boundary layer of moist air around the plant.

Other adaptations help to anchor the plant in the substrate, such as rhizomes that grow deep into a crevice or sandy soil, preventing a sudden river of rain from pulling the plant free. Burying their rhizomes and roots deep in the rocky gaps not only anchors the plant but also protects it from extremes of temperature. If the leaves die off from heat or cold, they can resprout from the protected rhizome. Rhizomes are also important pantries in these limited environments, feeding the plant when nutrients or water are otherwise scarce, or when the leaves have died back and are no longer producing sugars through photosynthesis.

Cracks in the rock hold only small pockets of soil and nutrients. To make the most of every crack, some plants (such as *Brachyscome* spp.; page 128) grow long stems or stolons that creep over the rocks until they find a new crevice containing soil and leaf litter.

Nepeta

These stems then produce roots, which grow into the crack, exploiting the resources while also creating another anchor for the plant. Other plants are even more amazing. *Cymbalaria muralis* (ivy-leaved toadflax; see page 24) very deliberately buries its own seeds in cracks higher up the rocky wall, allowing the plant to climb to where there will be less competition for light.

A group of plants called lithophytes grow on rocks, facing very similar challenges to epiphytes (plants that grow on other plants; see Chapter 7), and many species that grow on trees will also grow on rocks in the same environment. They use glues exuded from their roots to grip the surface they are climbing or hanging from, and take in moisture from their humid environments.

Sempervivum tectorum

Making the most of tight spaces

While we typically think of plants growing in crevices in sun-baked rock gardens, such microsites are not always large, open spaces in full sun. Balconies, patio planters and hanging pots are, by their nature, similarly tight spots. The plants have a limited space in which to grow, and they may be exposed to similar extremes, depending on their position. Alpine or coastal plants naturally adapted to these extremes can make wonderful pot plants and require far less water or nutrients than many others.

To make the most of these species, ensure the potting mix is sandy and gritty to make it very free-draining. You could place interestingly shaped rocks to make mini-landscapes, and fill the spaces with potting mix before choosing attractive alpines to position in the pockets of soil.

A roof garden on a shed or bin store is another tight space. In these hard-to-reach spots, it's ideal to choose plants that will spread slowly. Some will happily self-seed as annuals in colder climates, popping up in the spring. Those that spread by making roots on their stems or producing pups (mini replicas of themselves), such as houseleeks (left and page 130), will add to the naturalistic patchwork of vegetation.

Enjoying (wild)life

Rockeries provide a refuge for an array of wildlife. The uneven landscape creates not only microsites for different plants, but also sunny places in which butterflies, lizards and slowworms can warm themselves, small cracks for solitary bees to nest in, and shady spots to shelter shrews and other small animals. Some birds, such as thrushes, use the rocks to crack open snail shells, while the deeper crevices protect amphibians and reptiles from the cold of winter.

By adding water dishes for wildlife and creating a diversity of microsites to provide refuge for all manner of other bugs, we can encourage a thriving garden ecosystem. Not only will this maintain a healthy balance of pollinators, pests and their predators, but also it will create a garden in which we can enjoy life.

Hedera helix

Sisyrinchium species
Blue-eyed grasses

A tufted lawn of deep green foliage is topped with delightful purple flowers. The petals darken towards a bright yellow centre, making the flowers pop against their green backdrop. Blue-eyed grasses are members of the genus *Sisyrinchium*, and many are very well adapted to growing in small spaces thanks to underground rhizomes and good self-seeding from year to year.

The flowers of blue-eyed grasses close at night and open during the day, in a process called nyctinasty. Just like solar garden lights, which sense when it's getting dark before turning on, plants can respond to changes in light. During the day the available light contains a lot of red light and very little far-red light, but as the sun sets, this is reversed, with less and less red light compared to far-red. This ratio of red and far-red light is detected by plants using special pigments called phytochromes. When the phytochromes sense that night has fallen, water is triggered to leave cells at the base of the flowers, closing each bloom for the night. In the morning the red light increases again and water re-enters the cells, opening the flowers to the new day.

The purple *S. bellum* is native to the western US and is the basis of a range of new horticultural varieties. *S. striatum* is a taller yellow-flowering species native to the alpine grasslands of Chile and Argentina. Both species are becoming popular in rockeries and rooftop gardens owing to their ability to tolerate the various extremes that are typical of such environments.

Limonium species
Sea lavenders

Splashes of papery pink and purple 'flowers' stand tall above leafy green rosettes. Sea lavenders (also called statice) are species of *Limonium* that perform well in tough conditions of various kinds.

The papery splashes of colour are actually the calyces (protective flower sheaths), which remain for a long time after the tiny white flowers have finished (*L. bourgeaui* is shown here). These long-lasting, brightly coloured calyces also make *Limonium* excellent for dried-flower arrangements.

As might be expected from the common name, some of these species grow in coastal areas, and their adaptation to living in sandy, well-drained (and sometimes very limited) soil makes them ideal for tight spaces in rockeries or rooftop gardens. *L. sinuatum*, for example, is native to the lands around the Mediterranean Basin, where it is exposed to hot, dry summers and cool, wet winters.

Part of *Limonium*'s tolerance is down to the rhizome, which can store resources when conditions are good, for use during dry periods. Many species also have semi-succulent leaves, which can store water in their cells, and the leathery leaves of many are the result of thick cuticles that prevent unwanted water loss. These often fast-growing, water-storing plants are excellent in roof gardens for reducing rain runoff that might flood the local drainage system.

In addition to their survival properties, these plants are magnets for pollinating insects. In urban areas, including roof gardens, where space is limited, sea lavenders provide long-lasting colour while encouraging the flutter of butterflies and the hum of bees.

Nepeta species
Catmints

A sea of purple alive with the buzzing of bees lines the path through a rooftop garden. Catmints are beautiful yet tough garden friends for well-drained, sunny situations, and there are many species and varieties. They flower for long periods, attracting a range of bees and butterflies, and their leaves are a preferred food source for various caterpillars.

One species familiar to cat owners is catnip, *Nepeta cataria*. Cats find it irresistible thanks to a chemical, nepetalactone, that is made in its leaves. Used in small doses, catnip tends to have a relaxing effect on pet cats, although this does differ among individuals. Interestingly, nepetalactone is not only attractive to domestic cats, but also seems to affect species of wild cat, including leopards and cougars.

Many species of catmint make nepetalactone in different amounts and can be attractive to cats to varying degrees. This means that some care is required when you are establishing new catmint plants in the garden, if they are not to be nibbled at or even dug up. The process of potting on or planting out will damage the leaves and roots slightly, releasing nepetalactone into the air and attracting the neighbourhood cats. This is mainly a problem in small, newly planted catmints, so it's worth cutting the bottoms off empty plastic bottles and covering the new plants to limit the spread of nepetalactone in the air. The plant will stop releasing it after a few days, at which time the bottles can be removed and the plants should be safe.

Conversely, nepetalactone (in combination with other oils in the leaves) acts as a repellent for some insects, including mosquitoes and some aphids. The repellent mix produced by the plant is called *Nepeta* essential oil – Neo for short – and is being used to formulate safer, more natural insect repellents.

Catmints are native across Europe, Asia and Africa, and there are species adapted to just about any condition, but most are very drought-tolerant and prefer well-drained soil. As we have seen, plants with silvery-grey foliage are particularly tolerant of drought, since their foliage reflects excess light, reducing the need for the plant to transpire to stay cool. The hairs on the leaves not only include essential oil glands, but also help to create a humid layer of air near the leaf surface. This means that when the stomata are open for gas exchange, water can both exit and enter the leaf by diffusion.

Drought-tolerance is one characteristic that makes catmints great in tight spots or roof gardens, but many types also spread gently using runners. These low-growing stems creep along in search of a new crevice, and when they find a patch of soil, new roots grow. This not only helps the plant to find water, but also reduces runoff from roof gardens.

Origanum vulgare
Oregano

Above a carpet of pinky-purple flowers, butterflies hover, vying for space with the many bees. Oregano, sometimes known as wild marjoram, is a resilient plant that forms a loose carpet or small mounds. An edible herb all year round, it's a hive of pollinator activity when in flower.

In common with many drought-tolerant plants, oregano produces smaller leaves when water starts to run short in late summer. Each leaf is covered in tiny hairs that trap moisture near the leaf surface. This boundary layer of higher humidity reduces water loss and also, it has been suggested, protects the plant against frost in exposed rocky sites.

Other species of *Origanum* are also drought-tolerant and well adapted to rocky, well-drained positions, and many are edible herbs. *O. majorana* (marjoram) and *O. onites* (pot marjoram or Greek oregano) are two commonly grown herbs that come in a range of varieties with different foliage and flower colours.

Preferring well-drained, sunny positions, oregano is able to adjust its rate of transpiration quickly according to the conditions, which means that after a long dry spell it can quickly resume water uptake driven by faster transpiration. This makes it effective at reducing storm-water runoff and an excellent addition in more extreme environments. Tough, attractive, edible and a magnet for pollinators, oregano certainly ticks all the boxes.

Craspedia and *Pycnosorus* species

Billy buttons

Cheery yellow buttons sway in the breeze above a rocky mountain-top lawn of silvery foliage. Billy buttons are native to coastal and alpine areas of Australia and New Zealand. The common name encompasses species from the genera *Craspedia* and *Pycnosorus*, most often found in well-drained or dry locations.

While billy buttons may suffer in constantly wet soils, they are otherwise very adaptable, suited to rockeries and rooftop gardens. Their silvery foliage reflects excess light and heat, reducing the need for transpiration to keep the leaves cool. The fine hairs that cover the leaves reduce water loss by making a humid boundary layer, and the plants form grassy clumps or tight, flat, low-growing rosettes, protecting them from wind exposure and reducing unnecessary water loss still more. This habit also helps the plant to retain rainwater, which the thick roots quickly absorb and store for use during dry periods.

While most billy buttons are perennial, alpine species are more likely than others to be frost-tolerant. Still, even if the cold restricts them, they self-seed in the spring and can grow as annuals. The sunny flowers are also excellent in cut-flower arrangements, and they keep their colour and form when dried, making them a delight the whole year round.

Brachyscome species
Swan river daisies/cut-leaf daisies

A thousand purple daisy flowers with sunny yellow centres spill out from the rocky crevice. The lovely fine, highly divided foliage is obscured almost completely by the mass of flowers. *Brachyscome* species (sometimes referred to incorrectly as *Brachycome*, with the 's' missing) are mostly native to Australia and many (including *B. iberidifolia*, shown here) are adapted to dry or alpine conditions, making them ideal for well-drained pots, rocky niches or rooftop gardens.

In much the same way as strawberries, many of these beautiful plants spread along the ground via stems that form runners called stolons. Each stolon can sprout new roots and shoots, and these can develop into a whole new plant that can take advantage of a local patch of resources. This also makes the plant easy to propagate through cuttings or layering to fill out a roof garden or sandy border quickly.

Some species of *Brachyscome* spread via underground rhizomes. Safely buffered below ground from temperature changes, the rhizome can also store resources between rain events and aid resprouting after days of extreme heat or particularly hard frosts.

While their fine foliage helps them to resist losing water, in exposed sites they also respond well to their neighbours. Huddled together, there is more protection from wind or frost, allowing taller, more vigorous growth.

These adaptations make *Brachyscome* species perfect for well-drained tight spots, and their prolific flowering in pinks, blues and purples attract a range of pollinators to provide endless joy to the gardener.

Aurinia saxatilis
Basket of gold

Gold spills out from a crack in the wall of a derelict building. Appropriately known as basket of gold, *Aurinia saxatilis* is remarkable in its ability to grow out of apparently solid rock.

The fibrous roots of basket of gold can find their way into the smallest crack, attaching the plant to the wall and finding any nutrients and water that might have collected in the crevices. It is also happy on chalky soil, and, indeed, can adapt to many situations, so as long as there is good drainage, it will grow well in more luxurious garden sites. The hairy blue-green foliage helps the plant to reduce water loss and reflect excess heat, while adding to its attractive appearance.

Basket of gold fills the air with a floral scent that attracts a wide range of pollinators. While scent molecules that waft through the air (often called volatiles) attract pollinators, other volatiles attract the creatures that feed on insect herbivores. When plants are damaged or chewed on by herbivores, the leaf cells break, releasing particular volatile molecules into the atmosphere. These tell predators, such as ladybirds (ladybugs) or parasitoid wasps, that there's a feast nearby. *A. saxatilis* is a member of the mustard family (Brassicaceae), of which many produce these volatiles and also make within their tissues compounds that are toxic to predators. This means that basket of gold is a delightfully attractive, tough, resilient plant that is able to grow in the tightest spots.

Sempervivum species
Houseleeks

A mesmerizing mat of spiralling foliage forms in a dazzling array of colours from greens through to reds and purples. Unsurprisingly, species of *Sempervivum* are popular succulents, being easy to grow (and ignore) as pot plants on windowsills or sunny balconies.

Native to rocky, mountainous regions of Europe and the Middle East, houseleeks have fleshy leaves that can store water for use during dry periods. In common with other drought-adapted succulents, they use CAM photosynthesis (see page 28), opening their stomata at night to capture carbon dioxide, which is then stored as a four-carbon molecule until daytime, when the stomata close to conserve water.

Combined with the frost-tolerance of many houseleek species, the drought adaptations of these plants makes them very well suited to small crevices in rock walls or shallow rooftop gardens, where conditions can swing between extremes. Their toughness is even reflected in their Latin name, *Sempervivum* – from *semper* (always) and *vivere* (to live) – so these plants are 'always living'.

Houseleeks produce pups (tiny versions of themselves) from the main stem (as seen on *S. tectorum*, shown here), and it is this habit that leads to the lovely mats of many spiralling rosettes. When several species are grown together, a vibrant, multicoloured patchwork results. Not only do the species come in many colours, but also some have fascinating features, such as *S. arachnoideum* with its spiderweb of woven hairs. The flowers, while often not the main feature, are delicate and starlike in shape. No matter the size of the garden, houseleeks will flourish and offer year-round interest from any well-drained nook, cranny or shallow pot.

Vinca minor
Lesser periwinkle

Glossy green leaves cascade over an embankment studded with lavender-coloured flowers. Growing in small spaces is not always about rooftops and rock crevices, and *Vinca minor* – which comes in various shades of purple, as well as pinks and white – is not only excellent for trailing over walls and roof gardens, but also wonderful tumbling from hanging pots.

While fairly tolerant of drought, the lesser periwinkle prefers some shade and will cover the ground even in the darkest corner. Its stems can sprout new roots when conditions suit, helping it to create a weed-suppressing mat and maximizing access to water when rain is scarce.

Take care when growing lesser periwinkle in a frost-free climate or in otherwise more favourable conditions, since this vigorous grower can become a weed; in parts of North America, for example, it is considered invasive, so – as always – check the local regulations. In a small garden, consider planting a variegated cultivar (such as 'Variegata' or 'Illumination'). These are usually less vigorous because their leaves, edged with white or yellow, contain less green chlorophyll and so make fewer sugars for growth.

Periwinkles are hardy down to about -15°C (5°F). If the winters are reliably colder than that in your area, consider growing the plant in hanging pots so that you can move it under cover during the colder months.

Hedera helix
English ivy

A green wall is encased in English ivy. The lobed juvenile leaves, tipped with white or gold, transition higher up to smoother-edged mature leaves. The branched flowers that are so popular with pollinators in late summer and autumn are followed by black berries, a staple for birds in the winter.

When space is at its tightest, the best option is to grow upwards. Ivy is remarkable in its ability to climb even the smoothest wall. It does this using clusters of adventitious roots (those that sprout from the stems). In English ivy, these aerial roots remain fairly short, growing root hairs that secrete glues to provide the support for climbing. As these root hairs dry out, they also curl, gripping bark, brick and other rough surfaces in much the same way as the hooks of Velcro.

This skill in climbing has given ivy a bad name for damaging walls and weighing down gutters, but research led by Tijana Blanusa of the Royal Horticultural Society has shown that relatively simple management practices can help. Using anti-graffiti paint or zinc or copper sheets and meshes around window frames reduces risk because the roots are less able to stick to these surfaces. The RHS's research field looks like a series of chimneys arising from a lawn, as if little communities of houses live underground. These 'chimneys' covered with combinations of ivy and other climbers have shown that ivy is valuable not only for increasing insect biodiversity, but also for reducing humidity and damp around buildings. Ivy leaves are also excellent at absorbing air pollution.

The leaves that remove pollution from the air come in two different shapes, depending on the stage of development, making this a heterophyllous species. The juvenile leaves are the familiar ivy-leaf shape, with five lobes (this leaf type is called palmate). These lobes disappear in the mature leaves, resulting in smoother cordate (heart-shaped) leaves with a pointed tip. Both juvenile and mature leaves can be found on different parts of the same plant, and the flowers and berries develop on the more mature sections.

A native of Europe, English ivy is vigorous and can become invasive, but there are plenty of less vigorous cultivars, including some that remain in their juvenile growth form. While this means they cannot provide flowers and berries for wildlife, it does make them less likely to spread. As with most variegated cultivars (see page 131), variegated ivies are less rampant. Blanusa also tells me about such species as *Hedera hibernica*, which look like English ivy but are less vigorous and may be more suited to some areas. Still hesitant? There are also varieties without climbing roots, and these are great for ground cover or trailing from hanging pots.

While some varieties of English ivy have been selected for practical reasons, such as minimizing climbing or invasive potential, they also come in many different leaf shapes and colours. In greens, silvery greys and golds, some are plain and some variegated, and some tipped with gold or white, while others have hearts of gold. There are even varieties with frilly edges.

Rooftop gardens

A meadow stretches in front of me, its tufts of colourful prairie flowers contrasting with the patchwork of green and silver foliage. From my vantage point, it appears to reach into the distance and blend into the tree canopy beyond. But in fact, this naturalistic vista is on the roof of a building with just 20cm (8in) of soil, and is the work of the garden designer Denis Dujardin.

Dujardin explains that he works on roof gardens with a bioengineer and architects. Most 'flat' roofs actually slope very slightly, so that water drains towards one corner, leaving the rest very dry, very quickly. To avoid this, when working on a new building, Dujardin encourages the architects to design the roof perfectly level. Then, when the plants are put in, they all have equal access to water, resulting in these lovely meadows of flowers.

Rooftops pose challenges for plants, even in planters. Most such spaces are in full sun all day and exposed to the wind, so they can experience extreme swings in temperature, and go from pouring rain one day to baking sun the next. As Dujardin lists his favourite plants for these rooftop gardens, colours swirl through the air. The yellows of *Oenothera macrocarpa*, *Genista lydia*, *Sisyrinchium* spp. and *Sedum takesimense* give way to the fiery hues of *Kniphofia* spp. (red-hot pokers) and the oranges of *Pilosella aurantiaca*. Purple heads of *Knautia arvensis* float above *Nepeta* 'Junior Walker', *Pulsatilla* and *Caryopteris*, and blend into shades of pink thyme, *Dianthus carthusianorum* and the intensely pink-red *Knautia macedonica*.

But these flower colours are not the only interest. Texture is also important, and the bright pink flowers of *Lychnis coronaria* top grey-green leaves next to the fluffy silver foliage of *Stachys byzantina*, contrasting further with the strikingly sharp appearance of the grey *Euphorbia myrsinites*. Planted in drifts among this stunning array of flowers, grasses waft in the breeze, including *Stipa tenuissima* with its delicate, frothy-looking flowers and *Sesleria nitida* with short brushes standing above tufts of blue-green foliage.

Part of the success of the gardens Dujardin designs is his mixed planting strategy. His best tip is to work with nature, letting the plants self-seed gently and spread into the space. If you start with a mixed planting, some species will do better than others, outcompeting those that are not so well adapted and creating a unique patchwork.

I'm also interested to learn that one of the plants that grows well in rooftop gardens, *Knautia macedonica*, is often more showy, standing taller in the tough conditions than those grown in 'better' conditions, which flop without support. This is a good lesson for all gardeners. Pampering plants with fertilizer and water is often not the best way towards strong, resilient plants. Instead, tough love will produce the happiest plants that can withstand extremes of weather.

Phalaenopsis

Chapter 7

Humidity: An Unexpected Water Source

Introduction
Humidity

Enveloped by every shade of green, from the large leaves of the understorey to the moss-covered trunks and the epiphyte-covered branches above, mist hangs from the treetops, adding grey-blues to the palette. It's early morning, but already the sticky atmosphere hints at the heat of the day to come. Water dripping from the leaves beats a syncopated rhythm as it splashes its way to the forest floor. As the sun rises, moisture is everywhere.

Hot, moist tropical rainforests are often the first habitat to come to mind when we talk about humid climates. Many humid tropical plants are popular house plants (see page 154) or the holy grail of gardeners in cooler climates, pushing the winter boundaries for a tropical-inspired garden. However, there are humid habitats all the way from the tropics to temperate regions, and some even exist where rain is scarce.

Before we explore the amazing species that grow in these environments, let's find out about humidity, how it affects plants, and the adaptations they have made to humid climates.

Fuchsia boliviana

What is humidity?

Humidity is water vapour in the air. Moisture gets into the air through evaporation from oceans or lakes, or via plant transpiration (water exiting the leaves through stomata). Amazingly, in the tropics – where transpiration rates are the highest – more than double the amount of water in the entire atmosphere is transpired through leaf stomata every year. This shows how important plants are for increasing water in the atmosphere, to form clouds and eventually make its way back to the ground as rain.

Absolute and relative humidity

To understand humidity, we need to introduce a few terms. The first is absolute humidity (the amount of water in a volume of air), and is given as grams of water per cubic metre of air. Basically, absolute humidity tells us how much moisture is in the air.

But the maximum amount of moisture the air can hold changes with temperature. Warm air can hold more moisture than cold air. At 5°C (41°F) the air can hold just 7g per m^3. At 20°C (68°F), the most water it can hold is 18g per m^3, while at 30°C (86°F) it can hold 30g per m^3 of water. The amount of moisture in the air at that temperature is what drives evaporation, so if the air is holding 95 per cent of what it is able to hold, we feel sticky because our sweat can't evaporate into the almost saturated air. At 100 per cent or full saturation, rain or dew forms.

The value given in weather apps is called relative humidity: the amount of moisture in the air relative to the maximum moisture the air can hold at that temperature. Let's say it's 20°C (68°F) and the atmospheric moisture is 12g per m^3. We can look up the maximum moisture the air can hold at that temperature – 18g per m^3 – then divide 12 by 18 to get 0.67 and multiply by 100 to get the percentage. This gives us a relative humidity of 67 per cent. Helpfully, weather forecasters do this for us, so when the forecast is for 70 per cent humidity, the air is likely to contain 70 per cent of the moisture it could hold at the forecast temperature. Relative humidity is part of what many weather apps call the 'real-feel' temperature, because in low humidity, we feel cooler from evaporating sweat than we do when humidity is very high.

Byrsonima

Dew point

One more term relating to humidity that is of interest to gardeners is the dew point. This is the temperature at which the moisture in the air will begin to condense and form dew. Put another way, it is the temperature at which the relative humidity would be 100 per cent. For the example above, with a moisture content of 12g per m^3, we can use a dew-point calculator to tell us that just under 14°C is the temperature at which 12g per m³ would begin settling as dew.

Relative humidity and dew point control how much water plants lose through transpiration and how much a leaf might need to cool itself to encourage dew to settle on it. That is why it's important to understand both concepts when exploring plant adaptations.

Tillandsia

Hot and sweaty

One thing we all notice when relative humidity is high is that we feel hot and sticky. This is because the atmosphere is already holding almost as much moisture as it can. Instead of our sweat evaporating and removing heat as it does so, making us feel cooler, it stays on our skin, leaving us feeling hot and clammy. The same is true for plants. As plants metabolize, they create heat much as we do. When they become too hot, their leaves try to cool down through transpiration, but when relative humidity is high, cooling by transpiration is reduced and the leaves are at risk of overheating.

In environments of high heat and humidity, leaves create airflows over and around them to increase the chance that moisture will evaporate. They do this by developing holes. Air warmed by the leaf can slip through the holes, pulling slightly cooler air behind it in a process called convection. The Swiss cheese plant (page 146) is a classic example of a tropical plant with holey leaves that help it to stay cool.

Water from air

As we saw in Chapter 4, plants in mountain deserts can take up moisture directly from the air. This ability is not restricted to humid deserts, however, and it can be essential even in tropical rainforests, where competition for water and nutrients in the soil can be extremely high. Drip tips on the leaves of many forest plants funnel condensed water to the soil, where the roots are. Another common strategy is to bypass the soil altogether and take up water through the leaves.

Many warm, humid forests support a large number of epiphytes – plants living within the tree canopy. Life in the canopy is surprisingly extreme, with the availability of water dependent on unpredictable rain, and soil existing only as small patches of leaf litter caught in the branches.

To cope in this challenging environment, many epiphytes have adapted their aerial roots to capture and store water and nutrients like a sponge. To do this, they have an outer layer of cells that covers the whole root, with air spaces between the cells. This spongy layer is called the velamen and allows water (from condensation or rainfall) to be wicked along the root. Even more interestingly, while most soil-based roots have a short region just behind the root tip where nutrients are taken up, the aerial roots of epiphytes can absorb nutrients inside the spongy cell layer along their entire length.

The plants discussed in this chapter are native to humid environments. They will become of more use as the climate changes, bringing greater warmth and more rain. Understanding some of their adaptations can help us to grow tropical-inspired gardens or keep our house plants happy.

Monstera

Phalaenopsis species
Epiphytic orchids/moth orchids

Large flowers in a dizzying array of colours and patterns nod at the end of arching stems, and aerial roots emerge from between thick, dark green leaves to grow in all directions. *Phalaenopsis* or moth orchids are among the most common orchids to be grown as house plants, but in their native tropical forests they grow on trees, spending their whole life in the canopy.

As discussed on page 141, life in the canopy is surprisingly challenging. To make the most of brief spells of rain, the moth orchids' aerial roots soak up water and nutrients and hold it within the cells until it can be taken into the core of the root.

When keeping epiphytic orchids, such as *Phalaenopsis*, don't be tempted to water them on a schedule, but rather water only when the roots turn from green to brown. These popular plants are used to short, intense bursts of rainfall and high humidity in their native forests, so frequent misting and a short soak once in a while will keep them happy. Water them generously to refill the spongy layers, then allow to drain so that the plant is not left sitting in water.

Schlumbergera species
Christmas cactus/holiday cactus

Bright red flowers resembling the dresses of flamenco dancers hang as beautiful pendants at the ends of arching green stems. Named after the time at which it flowers in the northern hemisphere, the Christmas cactus is actually native to the humid forests of Brazil, where it grows on trees as an epiphyte. *Schlumbergera* is similar to other members of the family Cactaceae in having modified stems rather than leaves for carrying out photosynthesis. Each stem is made of individual segments, creating the appearance of leaves.

The beautiful flowers of the plants in this small genus have led to the development of a range of hybrids in a rainbow of colours from reds, purples and pinks to yellows and whites. Christmas cacti are short-day plants, meaning that for flowering to occur they need a night length (or period of darkness) of more than 12 hours for about 4–6 weeks. Growing them as house plants in a room where the lights are on in the evenings will prevent them from flowering, so keep them in a less frequently used space and show them off in well-used rooms for short periods.

A great aspect of these species is that the stem segments can be broken off and placed in gritty, free-draining potting mix to make new plants. Mature plants sometimes even grow long, wispy roots from the joints between segments, probably to help them to take up water from the air.

Happiest in indirect light and able to tolerate very light frosts in otherwise mild climates, these beautiful plants will brighten the short days of winter.

Tillandsia species

Air plants

Seeming to grow from thin air, *Tillandsia* species are members of the family Bromeliaceae (the bromeliads). Living their lives in the canopy as epiphytes, bromeliads have developed fascinating ways to survive this challenging environment.

Without roots in the soil, air plants – in common with other bromeliads – are dependent on short, intense bursts of rainfall. Some 'tank' bromeliads have developed a leaf arrangement that captures and holds pools of water that become entire ecosystems. Insects that live in the water can be eaten by vertebrates, such as frogs and their tadpoles, which in turn defecate in the water, providing nutrients for the plant.

But nutrients and water are taken up by roots, right? Actually, many plants can take up both through their leaves. Air plants have special scales called trichomes on the surface of their leaves, and these act as valves for water and nutrient uptake. These scales are actually modified hairs and look like upside-down umbrellas. Each has a core made up of two cells called 'foot cells' and a 'dome cell' that sits above the foot cells, and wing cells. The foot cells are connected in the outer layer of leaf surface cells (the epidermis), while the wing cells sit above the leaf surface. Water on the leaf is wicked in under the wings and funnelled into the foot cells.

What's even more amazing is that the materials of which the trichomes are made change how well water moves through them depending on how wet they are. When the foot cells are filled with water, the membrane around each cell becomes more permeable, meaning that water can flow through into the leaf. When they dry out, the membrane becomes less able to transfer water, so moisture is not lost from the leaf.

These trichomes can be seen through any basic microscope using a technique called an epidermal peel. To try this, paint clear nail varnish on the surface of an air-plant leaf. When it is dry, put some sticky tape over the nail varnish and carefully peel it off the leaf; it should bring the nail varnish with it. Now stick the tape to a glass slide and view it through a microscope.

These valves are one adaptation that helps air plants and other bromeliads to make the most of water condensing from the air on their leaves, or bursts of rain, but many species also have other strategies in common with xerophytes, which we saw in Chapter 4. One of these is CAM photosynthesis, the ability to trap carbon dioxide at night and attach it to an acidic molecule stored in the vacuole (fluid-filled cell cavity) until daytime, when it is released to make sugars with the help of energy from the sun. This method means the stomata can be closed during the day, preventing water loss.

With their stunning flowers and often attractive leaves, these fascinating plants grow well in a humid environment where the temperature stays above freezing. Try them in that most traditional of air-plant growing spots: the bathroom.

Monstera species
Swiss cheese plants

Giant leaves are punctured with holes like a Swiss cheese. *Monstera* species are among the most commonly grown house plants in the world. Native to the humid tropical forests of South America, they can begin life in the ground before climbing the nearest tree into the canopy, where they will spend the rest of their lives.

As the plant grows upwards, its leaves typically increase in size and change shape, beginning to develop the holes for which the mature leaves are recognized. It also produces roots from the stem as it matures. One type of root that the Swiss cheese plant grows is for clinging on, as we saw in the English ivy (page 132). These clinging roots are short and grow sideways around the tree trunk, exuding glues to help the plant climb.

The other type of root *Monstera* grows from its stems is much longer and grows towards the ground in search of nutrients and water. These 'feeder' roots also often have a layer of cells that act as a sponge (velamen), trapping the water until it can be taken into the root.

These aerial roots are really important for supplying water to the increasingly large leaf area, which requires far more water than the original stem and roots can provide. They can even let the original roots die off, severing their connection to the ground and leaving them free to roam the canopy.

Alocasia species
Elephant's ears

With its leaves that come in interesting shapes and colours, *Alocasia* species are popular as house plants and create a tropical feel in gardens in mild areas. Native to tropical and subtropical Asia and eastern Australia, they are shade-loving plants in the aroid family (Araceae). Where the temperature stays above freezing, they are surprisingly tough plants, producing large underground rhizomes that store resources so that the leaves can die back when conditions are tough and resprout when things improve.

Although they are not usually the focal point, the flowers of *Alocasia* are interesting (shown here in *A. micholitziana*). In common with those of *Symplocarpus* (page 70), they generate heat to dissipate their scent and attract pollinating flies. The flies gather near the base of the flower, where the female parts ripen first. The flies mate and lay eggs, rubbing the pollen they collected from a previous flower all over the anthers, thus pollinating the flower. As the upper, male part of the flower ripens, the flies are encouraged upwards through a narrow entryway. There they inadvertently collect pollen from the stamens before flying off to find the next flower.

This ingenious strategy prevents the flies from pollinating the flower with its own pollen, ensuring cross-pollination. Not only visually attractive to us, *Alocasia* can also turn on the heat to be irresistible to pollinators.

Musa species
Bananas

The very picture of tropical vegetation, *Musa* species – including the somewhat frost-tolerant *M. basjoo* – produce arching leaves that are often several metres long. There are many varieties with a range of leaf colours, from dark purple through to greens and even variegated options.

Musa species are monocots and thus more closely related to grasses than to trees. Their 'trunks' are actually quite similar to grass stems. The arching green end of the leaves gives way to lighter green leaf sheaths, which clump together to make the trunk. This arrangement is called a pseudostem. Below ground they form rhizomes from which new suckers can grow, resulting in clustered bunches or groves of these pseudostems.

While the leaves conjure images of the tropics, the flowers are just bizarre. Also called banana hearts, they are encased in several layers of colourful bracts. As each outer layer of bracts opens, female flowers are revealed and, when pollinated, make the rows of banana fingers. Then the next layer of bracts unfolds with another row of female flowers. Eventually the female flowers give way to the male flowers, which are enclosed in the bracts at the end of the inflorescence (flower stem). Often deep red or purple, the bracts create a stunning contrast with the bright yellow stamens as they are uncovered.

For a garden with a tropical feel, it's hard to bypass the banana. Choose a variety with a hardiness suited to your local area, and if you must grow it in a very cold climate, wrap it in the autumn with garden fleece or a similar insulating material until the worst of the winter weather is over.

Arecaceae
Palm family

A vision of white sandy beaches, turquoise ocean and deep blue skies is interrupted only by the presence of palm trees. Inland from the dunes, the rainforest of this tropical paradise is populated by palms of all sizes, from small understorey species to climbers and tall trees reaching into the canopy, exploding with fan-shaped leaves as they grow towards the light.

Palms have two main leaf shapes. One radiates from a central point and is called palmate, like fingers radiating from the palm of the hand. The other is called pinnate, in which long leaflets branch off a central rib – much like a feather. These leaves can be really large. In fact, the raffia palms (*Raphia* spp.) have the longest leaves of any plant, with pinnate fronds up to 25m (82ft) long.

In warm, humid climates where water is not limiting, large leaves can be an advantage for capturing light in the understorey or encouraging rain to drip around the roots. However, they heat up more than small leaves, so greater transpiration is needed to keep them cool. That is why we typically find these large-leaved plants in warm, wet climates. As ever, there are exceptions to this; date palms, for example, are fairly drought-tolerant and are grown throughout the arid and semi-arid parts of the world.

For those in cool climates who want that tropical feel, all is not lost. Palm species that are hardy down to -15°C (5°F) include *Chamaerops humilis*, *Trachycarpus fortunei* (shown here) and *Rhapidophyllum hystrix*.

Fuchsia species
Fuchsias

Clusters of bright red tubular flowers hang below dark green leaves, giving way later to edible dark purple fruit. Growing into a large, lush shrub, *Fuchsia boliviana* (shown here) is an attractive addition to gardens in mild climates. In common with other natives of the cloud forests of South America, where it originates, it has leaves with pointed tips that encourage moisture condensing on the leaf surface to drip off, feeding the roots below.

The fuchsias widely recognized in horticulture are often hybrids of species from the same region, some of which have dramatic two-toned flowers, sometimes in wildly contrasting colours. The bright hues of fuchsias attract hummingbirds, which pollinate the flowers in their native lands.

The members of another group of fuchsias from central America have small leaves, a trait that is reflected in the scientific name, *F. microphylla* (*micro* = small; *phyllon* = leaf). The subspecies and varieties in this group produce tiny flowers, bursts of bright colour that mature to edible dark berries.

While the huge range of horticultural hybrids and varieties of fuchsia include large, blowsy ornamental flowers that are delightful in cottage-style gardens, the more delicate species, including *F. boliviana*, give the feel of a lush tropical forest. Preferring to keep cool in shady positions during the heat of summer, they need protection in temperatures below -5°C (23°F). Once established, they can tolerate short periods of drought – perfect for that summer jungle oasis.

Byrsonima species
Locustberry/nance

Candle-like flowers flicker in shades from yellow to dark orange, standing tall above thick, glossy green leaves. The flowers of these tough, attractive shrubs or small trees are followed by edible yellow fruit.

Locustberries, such as *Byrsonima lucida* and *B. sericea*, and nance (*B. crassifolia*) are slow-growing but very adaptable trees that are native to the tropics. Able to cope with varied environments, from exposed dunes to tropical forests, they can change their growth habit to help them survive. Dune environments tend to be very dry, with rain draining away through the sand very quickly, and wind that increases evaporation. To overcome this, species of *Byrsonima* growing in dunes produce greater numbers of smaller vessels for water transport than those that grow in wetter environments. This helps to prevent a process called embolism, whereby air bubbles get into the vessels, stopping the movement of water. By having many shorter vessels, an embolism can simply be bypassed using another vessel.

Another adaptation found in *Byrsonima* growing in dry environments is small leaves, which reduce transpiration. These leaves are also thick, with closely packed cells, which increases the effectiveness with which they can fix carbon dioxide. Being more efficient means they can close their stomata to prevent water loss without reducing sugar production and therefore growth.

Dicksonia and *Cyathea* species
Tree ferns

Long, soft fronds unfurl from the centre of a furry trunk marked with the scars of previous fronds. Tree ferns, of which *Dicksonia antarctica* (shown opposite) is an impressive example, conjure images of misty rainforests, where they grow protected by the tree canopy.

Tree ferns are native to rainforests from temperate through to tropical regions. In common with many ferns, tree ferns produce rhizomes, but their rhizomes, surprisingly, grow vertically into the air like a tree trunk. It is from this trunk that the fronds unfurl. As each frond dies, it drops off, creating leaf scars or short stumps along the trunk, and these scars capture leaf litter and retain water. Each frond is ringed with short aerial roots, which absorb the nutrients and water trapped in the rough stem. As they mature, the aerial roots become woody, adding to the rough appearance of the trunk.

Being able to trap resources is valuable in a rainforest environment, where there is great competition for the nutrients and water in the soil. The root systems of densely growing trees and shrubs, all foraging in the same patch of ground, quickly deplete the substrate. This limiting environment leads to alternative strategies for the capture of resources, such as those of epiphytic orchids and air plants (pages 142 and 144). Tree ferns remain attached to the ground, but can capture and retain leaf litter and rain.

Another example was recently identified in a type of *Cyathea* tree fern that appears to have zombie leaves. In *C. rojasiana*, a native of Panama, the dead fronds remain connected to the trunk but hang down so that their ends touch the ground. Where they come into contact with the ground, these apparently dead fronds sprout roots that can take up water and nutrients. In the zombie leaves, the rachis (central spine of the fern frond) acts as a pipe, connecting the new (adventitious) roots to the trunk, while the pinnae (leaflets of a fern leaf) die off, releasing nutrients into the soil. The adventitious roots then absorb the nutrients released from the decaying pinnae and suck them back up into the growing tree fern. This is a great advantage in the competitive rainforest soils.

In the garden, tree ferns prefer shade and love rainfall running down the trunk, wetting all those little aerial roots. In the winter they will need protection from frost below -5°C (23°F), and in the spring new fronds unfurl from the centre of the top of the trunk, producing an umbrella of soft green foliage. Tree ferns can also grow happily in pots, as long as they receive adequate water. While still small they can be elevated on a pot stand hidden among other plants, to give the illusion of a larger specimen and create the rainforest atmosphere for which these lovely ferns are known.

Bringing the jungle indoors

Looking up, I see a lush green canopy. There is the sound of running water, and roots dangle like a curtain from fascinating tropical plants. Colourful foliage provides as much interest as the blooms of the orchids clinging to aerial structures. No, I'm not in a tropical rainforest but inside a temperate building filled with 'house' plants.

Those of us who live in temperate climates think of these plants as 'indoor' plants, but of course they did not evolve inside buildings, instead typically coming from subtropical or tropical forests. It's exciting to see the world of house plants blooming as a gardening style in its own right, with people becoming more aware of the advantages they bring to our mental health, our concentration and the air quality inside our homes and offices.

House plants can be tricky, however. For the best chance of success with each species, learn about where it is from. Is it an epiphyte growing high in the forest canopy, where it gets lots of light? Or an understorey plant that is used to low light? Then find the position in your home that matches that environment most closely.

There are two mistakes to avoid. The first is thinking a windowsill is the best growing position for any plant, and the second is overwatering.

While many plants need decent light, most don't prefer to be on a windowsill, against the glass, in direct sun. This position heats up to extremes (and gets colder than the rest of the room in winter). Instead, place your plants back a little, so that air can move around them, keeping the temperature stable. Also think about where the central heating is, since few plants like to be in the dry air directly above heaters or radiators.

Loving plants to death by overwatering is a very common problem. The optimum frequency of watering will depend on the size of the plant and that of the pot in which it's growing. Most house plants prefer a soak once in a while rather than frequent topping up. If in doubt about whether it's time to water, stick a finger about 5cm (2in) into the soil. If it feels moist, leave it; if it's dry, time to water. Ideally, put the pot in the sink or a bucket of water, soak it for ten minutes, then pull out the plug or tip out the water and allow the pot to drain before returning it to its place. If the pot is too large for this, position it in a tray and remove any water that is left in the tray after watering. If the pot is too heavy to lift, drape a cloth from the tray into a second tray to wick the water across.

Many house plants produce aerial roots. These can be misted to help the plant provide moisture to the increasing leaf area as it grows. The Swiss cheese plant in my office has huge leaves, which transpire lots of water, but is actually in a very small pot, so I spray the aerial roots with water most days and water it only when the leaves start to droop.

The aerial rooting nature of many of these plants makes them wonderful for propagating from cuttings. Cut below a node (where the leaf joins the stem) – many will have little bumps where the new aerial roots will grow out – then simply put the cutting in a jar of water until the roots appear. I've seen this done very attractively in clear glass vases or tubes – like cut flowers that will become a whole new plant.

A final note. While indoor gardening is becoming ever more popular, it's important to source plants responsibly. The poaching of plants from forests is destroying wild habitats, so buy yours from a reputable nursery and share cuttings among friends to avoid contributing to habitat destruction.

Hedychium gardnerianum

Chapter 8

In the Shadows: Making the Most of Every Sunfleck

Introduction
In the Shadows

Hosta

Frothy fern fronds bob in the gentle breeze. I look up into a dense, vibrant green canopy of foliage that creates welcome relief from the sun. Birds flit and dip into the birdbath in front of me. The hum of high-energy pollinator activity is muffled here, as if the softer light can take the sharp edges from sound as well as easing the contrasts of the midday summer sun. As my senses adjust to the lower light and softer sounds, I become aware of the delicate, fairy-like flowers of *Epimedium*, while the bench on which I sit provides the best vantage point for enjoying the groups of *Enkianthus* bells dangling below the foliage of the large shrub beside me. Many layers and shades of green produce fascinating shapes and structures, while the dappled light dances across the scene, constantly changing with the breeze and the progress of the sun.

While greens, blues, silvers and whites are the stars of this part of the garden, shade itself comes in different shades, from the darkest corner, where no direct light reaches and reflected light is minimal, to the front of the grove, where the early or late sun reaches its life-giving fingers into the foliage, providing the energy needed for growth. In the northern hemisphere, north-facing gardens might get no sun at all in the winter, but in the summer they can get many hours of direct light as the sun curves around the sky. In the southern hemisphere this is reversed, and the angle at which the sun can reach in depends on how close to the equator the garden is. Tall fences and buildings create solid shade against direct sun, but the areas underneath and beside them can still be light and airy, depending on how much light bounces off walls or windows.

These differences in shade also provide opportunities to grow plants that make the most of different microclimates. But how do shade-loving plants adapt to low light? Not surprisingly, many features of these plants are centred on maximizing the light captured for photosynthesis, so that's where we will start.

Fern

Seeing the light – and capturing it

One strategy that some plants use is to avoid shade altogether by growing up towards the light, climbing trees, fences or walls. Once they reach the higher light in the canopy, they usually behave in the same way as any plant growing in full sun.

Those that remain in shadow require different strategies. Some, such as those from the humid forests that we explored in Chapter 7, grow very large leaves so as to increase the chance of catching the light. Angled towards places where flecks of light are more likely to dance across them, the leaves do everything they can to maximize the light they encounter.

As part of making use of every fleck of light, many shade plants take steps to ensure that all the light reaching the surface is captured for photosynthesis. Without the energy obtained through photosynthesis, growth will slow, and if the levels continue to be low or drop further, the plant is at risk of dying.

The light that starts this process is captured in the green chlorophyll in the plant cells. To our eyes, most leaves look completely green, but at the microscopic level it can be seen that there are spaces between the green chlorophyll molecules, which light can pass through or bounce off. Shade-adapted plants pack in more chlorophyll to fill in some of those gaps and reduce the amount of light that escapes.

Once the light is captured, it proceeds as an energy parcel through a chain of actions that result in the production of sugar. Each of those actions occurs in specialized proteins. A ray of light must be passed to the first protein, which passes it to the next protein, and so on down the chain. If a protein in the chain is already holding a parcel of energy, it can't take another from behind it in the chain. If there are not enough of each type of protein to take the energy parcel, the energy is lost from the leaf rather than reaching the end of the chain. Think of a chain of people passing buckets of water to put out a fire. A person holding a bucket can't take a second bucket until they have passed on the first. If there are more people filling buckets from a constantly running tap than there are people passing buckets, the water from the tap will run away down the drain while those people are waiting to hand over their buckets, so it won't be put to use on the fire.

One solution is to have more people passing buckets, and this is similar to what some shade plants do. To avoid losing the energy from the light they can access, some produce extra numbers of proteins to keep the energy parcel moving down the chain. This means that once it has been captured by the extra chlorophyll, less light energy is wasted through traffic jams in the protein chains.

Epimedium

Sugar rush

As we watch the light filtering through the canopy and hitting the leaves in the understorey, those leaves are rapidly switching on processes that can increase the conversion of carbon dioxide into sugars. This is often one of the slower steps in photosynthesis, and when light is constant, this fact is less important, since energy will keep coming down the chain of proteins. However, light that flickers through leaves or is otherwise restricted may provide illumination for only minutes or even seconds, so these shaded leaves must be able to speed up the fixation of carbon dioxide.

To avoid losing the energy before carbon dioxide is fixed, some shade plants keep close at hand a large pool of the components that are needed. It's a little like mixing quick cement to hold up a fence post: if the pole is not in a pre-dug hole before the water is added to the cement, the cement will harden before it can be used. In the same way, if the molecules needed to fix carbon dioxide are not ready when the light energy comes, the energy will dissipate and no sugars will be made.

Plants for all types of shade

Light (or lack of it) is not the only consideration when choosing plants for a shady spot. As with anywhere else in the garden, moisture is also important. Some shaded positions may have surprisingly dry soil, despite feeling cool, especially where the roots of trees and larger shrubs quickly soak up rain. Other locations may be along a creek or near a pond that remains boggy in all but the driest summers.

Whether the conditions are wet or dry, there is a surprising range of shade-loving plants that will delight all tastes. This chapter lists just a few plant groups as examples, but the understoreys of woodlands and forests all over the world contain a wide diversity of species, all happily competing for those flecks of light. Shady gardens are the perfect habitat for many of these species, so look to nearby woods for inspiration, then head to a local supplier to fulfil your woodland dreams.

Enkianthus campanulatus

Ferns

Baskets of green fronds accented by unfurling fiddleheads thrive quietly in a shady corner of the garden. Ferns are a diverse group of plants from many families, several of which will grow happily in shade.

The diversity of this plant group is reflected in the wide range of frond shapes, sizes and arrangements, from single tongues arranged as baskets (the crow's-nest fern, *Asplenium australasicum*) or emerging seemingly at random (the hart's-tongue fern, *A. scolopendrium*), to fronds divided into rows of leaflets either side of the midrib, as in the Christmas fern (*Polystichum acrostichoides*) and the holly fern (*Cyrtomium falcatum*). In other species, each of those leaflets is further divided, creating a fine, lacy appearance, as in golden tassel ferns (*P. polyblepharum*). Maidenhair ferns (*Adiantum* spp.) have delicate, sometimes rounded leaflets accented by black stems, while the delicate fern (*Onychium japonicum*) has fine foliage that resembles carrot tops. There is also a range of ferns that grow as epiphytes (see page 141) high in the canopy.

As experts at surviving in shade, many ferns grow quickly despite the low light. One way they do this is by increasing the chlorophyll in their foliage, to make the leaves better at collecting light for photosynthesis.

Ferns are also interesting because they have two separate bodies during their lifetime. The fronds we see most often combine with rhizomes to make up one of these bodies, the 'sporophyte', so called because this is the version of the plant that produces the spores (from the Greek *spora* = seed and *phyton* = plant). In ferns, the spores grow to produce a different plant body, the thallus, which is small and looks like a wet lichen. The thallus produces sperm and eggs, which come together for fertilization to produce the sporophyte. The thallus stage of the fern's life cycle can often be seen on rocks in damp corners of a fern-filled garden, or along the sides of freshwater canals or ponds.

While we often think of ferns for shady, wet locations, several are actually very tolerant of dry shade. These include many polypody ferns (such as *Polypodium vulgare*), shield ferns (*Polystichum* spp.), hart's-tongue ferns and species of *Dryopteris*, such as *D. erythrosora*, the last of which produces lovely bronze foliage in the spring. So, whether you are looking for a shade-loving plant for indoors or outdoors, either large or small, evergreen or deciduous, in damp or dry conditions, there will be a suitable fern with foliage to suit your taste.

Zingiberaceae
Ginger family

Beacons of colour erupt from lush tropical foliage. Members of the ginger family are mostly native to tropical and subtropical forests and produce a wide range of colourful, interesting, exotic-looking flowers.

While most gingers are adapted to warm climates, a few genera are native to cooler alpine regions, and these provide slightly hardier options so that cool gardens can benefit from their luscious tropical appearance. Some, such as *Hedychium gardnerianum* (shown opposite), will not tolerate freezing, but more cold-tolerant species to look out for are *Cautleya gracilis*, *C. spicata*, and species of *Roscoea* and *Hedychium*. These will die back after the first frost, but a good mulch at the end of the autumn will keep their rhizomes warm enough to survive the winter, ready to burst into life when the soil starts to warm up in the spring.

Being adapted to growing in the shady understorey of forests, gingers – in common with many plants in this chapter – are more efficient at capturing light than those growing in brighter environments. Along with chlorophyll, plants rely on a number of components to convert light into energy for making sugars, and gingers increase the number of these extra components. This is like adding extra lanes to a highway, and it means that when the sun comes through a gap in the tree canopy, ginger can harvest light without creating a traffic jam in the chain of steps leading to sugar production.

Too much light, though, will flood the system, damaging the machinery for photosynthesis. One way to protect these pathways from too much sun is to shade them – but how can every leaf on a plant shade itself? The answer is for the leaf to roll into a tube. This means that a very small surface area is exposed to direct sun, while the rest of the leaf is shaded. It is a short-term survival strategy, however, and shade-loving plants will begin to show sunburn on the leaves if they are exposed to too much light for long periods.

In addition to their lush, tropical beauty, many members of the family Zingiberaceae are edible, used as spices in everyday cooking. The ginger we buy in the shops is the rhizome of *Zingiber officinale*, while its cousin cardamom is made from the seeds of *Elettaria cardamomum*. Another relative, turmeric, comes from the dried rhizomes of *Curcuma longa*, used not just as a spice but also as a dye, the colour of which depends on the acidity of the dyeing solution.

Ginger rhizomes from the greengrocer or supermarket may sprout if they are planted, and the resulting plant, when kept in a warm spot out of direct sunlight, will bring a lush tropical feel into the home at very little cost. These definitely won't survive a cold winter, but they will grow inside throughout the year, their floral beacons of light and colour erupting from the shadows.

Hosta species
Plantain lilies/hostas

Spires of purple, pink or white flowers stand proud above mounds of lush green and white foliage. With leaf colours ranging from blue-grey through shades of green to yellow, including every combination of variegation, and attractive trumpet-shaped flowers, hostas provide interest in shady gardens throughout the growing season.

That very range of leaf colours and variegation, which is determined by the amount of chlorophyll and other pigments that can absorb extra energy, means that hostas vary in their tolerance of shade. In general, varieties with pale green or blue leaves are better adapted to deeper shade, and will be unhappy in too much sun. But the good news is that there are more than 7,000 cultivars, so there is bound to be one precisely adapted to the level of shade of any garden bed.

Hostas also have a rhizome that means the plant can remain dormant during cold winters, storing reserves in preparation for fast growth when temperatures rise. The rhizome also makes hostas fairly resilient to short periods of drought, and this is especially true of those with thick, waxy leaves, which reduce water loss.

Regardless of the choice of colour, humans are not the only ones to love these attractive leaves; they are notoriously a favourite of slugs. A reliable method of control will be needed to keep our slimy friends at bay, so that we can also enjoy these delightful plants.

***Podophyllum* species**
Podophyllum

Mounds of star-shaped leaves are mottled in brown and shades of green, like tortoiseshell. 'Spotty Dotty' is a cultivar of *Podophyllum versipelle*, a native of wooded areas of China that grows about 50cm (20in) tall. These fascinating plants are quite a talking point in the shady garden.

'Spotty Dotty' and other podophyllums, such as mayapples (*P. peltatum* and *P. hexandrum*, shown here), grow well in dappled shade and are hardy to about -10°C (14°F). The rhizome lies dormant underground during the winter, and the fascinating leaves sprout early in the spring, enlarging like mini umbrellas. Deep pink (as in *P. versipelle*), red (*P. pleianthum*) or white (*P. peltatum*) flowers hang from stems below the leaves. These are followed by red or yellow fruit, depending on the species.

The roots of podophyllums partner with specialized fungi called mycorrhizae, which search out nutrients in the soil and provide them to the plants. In return, the plants give the fungi some of the sugars they have made via photosynthesis. This interaction is beneficial to both partners and is called a symbiotic relationship.

Interestingly, podophyllum rhizomes also make a compound called podophyllotoxin, which can stop cell division from occurring. Because of this, it is used in humans to treat external warts, and a modified version of the molecule is used to fight tumours, both of which are the result of excessive cell division.

Epimedium species
Barrenwort/bishop's hat

Sprays of spidery flowers appear to dance in the air above dense mounds of foliage. *Epimedium* species produce flowers in a range of hues, from bright reds, pinks, yellows, oranges and purples to pastel shades and white.

Barrenwort is native to woodlands and prefers dappled shade and well-drained soil. Below ground, the winter-hardy rhizomes store resources. The young foliage is often a different colour from the mature leaves, and the colours change again in the autumn, with many varieties turning red, orange or gold. Some are also evergreen, maintaining their structure through the winter. Many of these lovely plants are surprisingly tolerant of dry periods, using the rhizome to store water for the growing leaves to draw on when rain is scarce.

New varieties appear every season with novel combinations of leaf and flower colour. This is partly because different species can cross-pollinate, readily creating new hybrids, each with a slightly different combination of traits. This large range of new varieties, some of which prefer lighter or deeper shade, and wetter or drier soil, includes options with large and small flowers in a rainbow of colours. For a shaded garden bed, the new varieties of *Epimedium* are worth investigating...and again next year...and the year after that!

Fatsia species
Fatsia

Spidery strings of silvery white edge the branching green veins on deeply lobed leaves. *Fatsia japonica* 'Spider's Web' is just one of a range of *Fatsia* varieties, from those with completely green, shiny foliage to variegated types with light and dark greens, and even almost completely white leaves.

Fatsias are tall shrubs with white, globe-shaped flowers held above the foliage. They are known in Japan as *yatsuda* (eight fingers) because of their lobed leaves. *F. japonica* (shown here) has wide 'fingers' and *F. polycarpa* narrow ones, so they are often planted side by side for structural contrast.

Fatsias prefer shade and, in common with most plants in this chapter, produce more chlorophyll in their shaded leaves, making them very efficient at absorbing light. Despite the jungle aesthetic of their large leaves, they are hardy down to about -15°C (5°F), making them an incredibly versatile option in temperate gardens.

A very interesting hybrid of fatsia originated in 1912 in the Lizé Frères tree nursery in Nantes, France, when breeders crossed *F. japonica* with English ivy (page 132). Fatsia is in the same family (Araliaceae) as ivy, and amazingly this extraordinary cross produced offspring, which was named × *Fatshedera lizei* (the cross coming first to indicate a hybrid between genera), combining the two genera into a new genus. The new plant, known commonly as fat-headed Lizzie, is shrubby, with five-lobed leaves (like ivy) somewhere between those of fatsia and ivy in size. New cultivars have since emerged with a range of variegations, and they are worth seeking out.

Hydrangea species
Hydrangeas

Large oak-shaped leaves, deep green in the summer and red in the autumn, form a tall, domed shrub topped with conical bunches of ice-cream-like flowers. *Hydrangea quercifolia* (shown here) is named for its oak-leaf-shaped foliage (oak being *Quercus*), and is a favourite for its year-round interest.

While many of us are familiar with *H. macrophylla*, its flowers blue on acidic soil and pink on alkaline, there are many other species and varieties that are worth exploring. Although their love of water is well known, hydrangeas have an underground rhizome that allows them to tolerate periodic drought in otherwise damp climates. Some are large shrubs, such as the varieties of *H. arborescens*, *H. aspera* and *H. quercifolia*, while others, such as *H. serrata*, are smaller.

The various species also produce varied blooms. Some have the recognizable mop-like flowers of *H. macrophylla*, either as dense balls or as tall cones made up of large-petalled blooms. Others have fine, lacy flowers at the top of the flower head, surrounded by a smattering of larger (usually infertile) flowers. *H. macrophylla* 'Mariesii Perfecta' is one such lace-cap type and grows up to 2m (6½ft) tall.

There are even climbing hydrangeas, such as *H. petiolaris*. This vigorous climber clings to walls using modified roots along its stems, similar to the way in which English ivy (page 132) climbs.

Whether you are looking for large flowers or interesting foliage on small or large shrubs, the range of modern varieties will provide delight in dappled shade.

Daphne odora
Daphne

Out on a cold morning walk in early spring, a sweet scent reaches me. On searching, I find the small pink flowers of an evergreen shrub, *Daphne odora*, to be the source.

Well known for their scents, daphnes are small, slow-growing shrubs, making them excellent in small gardens or towards the front of a large flower bed. They are not fond of long drought or waterlogged roots, so to keep them happy, plant them in a shady spot in well-draining soil. Wherever possible, position them close to a path so that their scent can be enjoyed in the winter months.

While many daphnes are evergreen, there is one species, the Japanese native *D. pseudomezereum*, that has an interesting adaptation to shade. It is summer-deciduous, meaning that it keeps its leaves through the winter and drops them in the summer. By contrast with plants that drop their leaves in order to reserve water during dry summers, this one drops its leaves when the forest canopy above closes at the beginning of the summer, reducing the light levels to such an extent that making sugars is no longer viable. The plant resprouts in autumn as the dense temperate forest loses its leaves, capturing light throughout the winter under a leafless canopy.

There are now many varieties of daphne to choose from. Most flower in late winter and early spring, bringing their lovely scent and evergreen structure to the fore when the rest of the garden is slumbering.

Enkianthus campanulatus
Enkianthus

Pink bell-shaped flowers hang in bunches amid vibrant green foliage from spring into summer. As the seasons turn, the green of the foliage gives way to oranges and deep, fiery reds.

Enkianthus campanulatus is an attractive, slow-growing large shrub. Like other members of the family Ericaceae, including heathers, rhododendrons, lingonberries and blueberries (with which it shares bell-shaped flowers), it prefers acidic soils. *E. campanulatus* is named in honour of these flowers, from the Latin *campanula* (little bell).

This is probably the most widespread species of *Enkianthus* in gardens, having been introduced to the UK in the early 19th century, and it also has the widest native range of the Japanese species, being found in open woodlands. Others include *E. perulatus*, which produces white bell flowers that are drawn in at the open end, giving them the appearance of the puffy short sleeves of a summer blouse. Another is *E. cernuus*, which has flowers in shades of pink. There are many varieties in intermediate colours ranging from white to deepish red, and some have blooms that gradate from almost white at the top to dark pink at the bottom.

Most species are cold-hardy and prefer dappled shade, although in wet, cool summers, some varieties can cope with direct sun. When grown in shade, they are also tolerant of dry periods. This adaptability, together with their year-round interest, makes them valuable garden plants that could be more used.

Camellia species
Camellia

Small white fairy skirts hang among pink buds from pendulous branches adorned with dark green leaves. *Camellia lutchuensis* var. *minutiflora* is certainly a less traditional camellia than most.

If the pendulous growth is not to your taste, *C. rosthorniana* also produces masses of small white flowers that open from pink buds, but it maintains a more upright shape. Other less traditional varieties include *C. sasanqua* 'Yuletide', which does indeed flower at Christmas (in the northern hemisphere), its flowers bright red with intensely yellow centres. Similar in size and shape, *C. s.* 'Versicolor' has lovely multicoloured flowers tipped with dark pink, fading to white, met by a circle of lemon-yellow stamens.

There are also modern varieties of the more familiar *C. japonica*, some of which have densely swirled petals and others amazing flecks of colour. My favourites, though, and the most ecologically valuable, are the single-petalled varieties, with their open stamens covered in pollen ready for pollinating insects and birds to find.

While they do tolerate full sun, many camellias much prefer dappled shade, and all require acidic soils. Shade-loving ones include the species from which the favourite brew of tea is harvested: *C. sinensis*. When grown in full sun, its leaves become bitter because of the extra molecules the light-stressed plants produce. Good-quality tea is actually grown in partial shade. So, whether your choice is a traditional or more modern variety, you can sit and enjoy it with a good cup of shade-grown *Camellia sinensis*.

Thriving in low light

From a world-leading rock garden, we walk through into a woodland area filled with shade-loving plants. The path wends its way through ferns, *Polygonatum* and *Trillium*.

Connor Smith, an expert from Utrecht University Botanic Gardens, tells me that when choosing plants for a shade garden it's important to consider water availability and determine the type of shade. A shade garden could be anything from a fast-draining rockery to a swampy forest with soil that is constantly wet. When does the rain come? Is there less rain in the summer, when the thirsty, fast-growing trees are soaking up all the ground moisture? The answers to these questions will help you to determine whether the position is damp or dry shade, and will inform your choice of the best species to fill out the planting.

Shade also comes in many intensities. How shady is it? Does that change with the seasons? Smith reminds me that deciduous trees let in lots of light in the winter and early spring, which can fuel spring flowers, such as snowdrops, anemones and bluebells. These plants make the most of the early spring sunshine, then remain dry and dormant throughout the warmer months. Such conditions are very different from the consistently shady nooks beneath evergreen trees and shrubs.

Other wonderful woodland plants include the trilliums. These plants are a celebration of threes. Each flower has three sepals (outside the petals), three petals, and whorls of three stamens. As if that weren't enough, they typically hold their flowers above a windmill of three leaves. A few examples are *T. chloropetalum*, with its dark red flowers, and *T. albidum*, with white flowers, both native to the western US. Smith tells me that the way to distinguish *albidum* from a white form of *T. chloropetalum* is that the flower of the former will smell like a fine rose. The seeds of these plants, in common with those of many woodland plants, are spread around the garden by ants.

Polygonatum odoratum and *P. verticillatum*, commonly called Solomon's seal, are two more delightful plants for dry shade. They are also happy along the brighter edges of a woodland garden.

For wet shade, Smith suggests species of *Arisaema*, such as the Middle Eastern and Asian native *A. flavum*. These have lush foliage and the snake-like flowers typical of the aroid family. All prefer wet positions and mild, maritime climates.

At the other extreme – dry, shady rock gardens – two options are species of *Ramonda* and *Haberlea*. These two groups of 'resurrection' plants (page 194) can make you think they're dead, but if you throw water on them, they bounce back. Some of the toughest plants around, they also have beautiful flowers.

As a parting tip, Smith explains that while many plants can grow in shade, some still need plenty of light if they are to flower. If something isn't flowering, the solution may be to try moving it to a brighter spot.

Odontarrhena serpyllifolium

Chapter 9

Extremophiles: The Toughest of the Tough

Introduction
Extremophiles

A white crust crunches underfoot. Salt. Right now it's a white salt pan, but from time to time this landscape fills with water, and with it an explosion of life. Kati Thanda–Lake Eyre is an ephemeral salt lake (meaning that it fills with water occasionally) in northeastern South Australia. When it is full, seeds that were lying dormant spring into life, creating a green oasis that attracts birds and other wildlife, triggering a breeding boom. But even when the lake is dry, there is still life in it. This salt-encrusted basin is surrounded by salt- and drought-tolerant plants that have adapted to cope with the conditions while making the most of every drop of water.

Salt lakes are not the only extreme environment on the planet. In this book we have explored many challenging places, from the watery to the seasonally dry, and from icy cold to hot and humid. Now we will explore the toughest of the tough: those adapted to the extremes, and often combinations of extremes.

Armeria maritima

Extremophiles

In fact, many plants in this book could be considered extremophiles. The term is applied loosely to any organism adapted to conditions that are normally considered challenging for regular growth and reproduction. Xerophytes – plants adapted to dry conditions, such as those we explored in Chapter 4 – could arguably also be called extremophiles. The most extreme of these are plants that can dry out their tissues, appearing dead and desiccated, but that come back to life when watered. Often referred to as resurrection plants, these desiccation-tolerant species are truly amazing (see page 194), and many are very attractive – when alive.

Lack of rainfall is not the only condition that makes water hard to get. Epiphytes growing in the tree canopy depend on short bursts of rain that provide lots of water, which is funnelled down the trees, leaving these plants with their minimal soil to wait for the next downpour. Between showers they may depend on absorbing humidity from the air through spongy cell layers, or wicking condensed moisture along fine hairs, carefully preserving it until the next opportunity arises. Freezing conditions also make water hard to come by, since plants cannot take up ice.

Having explored the role of changing ion concentrations to move water around in drought-tolerant plants (see page 80), it is perhaps not surprising that soil containing high levels of ions, such as salts or heavy metals, also make it difficult for plants to absorb water. A high concentration of salt in the soil tends to draw water out of the plant, so it is hard for the plant to access and hold on to it. Plants facing these conditions often have adaptations in common with those from dry landscapes. Salts and heavy metals are also toxic to cellular processes, so plants that are able to live in these conditions are definitely extreme.

We have discussed adaptations for dry and cold in previous chapters, so here we'll explore the adaptations found in halophytes (salt-loving plants) and metallophytes (metal-tolerant plants).

Halophytes

Salty habitats may be tough environments for plants, but they are surprisingly widespread. From coastal mangroves or salt marshes and salt lakes, to the sides of roads where salt is used to prevent ice from forming in the winter months, salt-loving plants are all around.

But why is salt so bad? The answer is that it can damage cell membranes and interrupt important cellular processes, and it affects the osmotic movement of water into and around plants. Water moves from areas of lower ions (salts) to areas where there are more ions. In non-saline soils, the roots contain more ions than the soil, so water tends to move

Asphodeline lutea

into the roots, helped along by transpiration pulling water through the plant. But when the soil is full of salt, moisture wants to move *out* of the roots. This also triggers a drought response, which shuts the stomata and halts transpiration. Salt-tolerant plants overcome this by accumulating ionic sugars and amino acids in their cells and tissues, thus modifying the number of ions. This helps them to absorb water from the soil and is used to move water into cell compartments or through to the xylem (circulatory vessels).

Many halophytes will also take up salt, but it must then be kept safely away from the membranes and prevented from interrupting cellular processes. One strategy is to store the salt out of the way, such as putting it in old leaves that are then shed, removing the salt from the plant.

Another strategy is to spit out the salt. While this can happen at the roots through specialized gates that pump out the salt, it can also happen through salt glands in the leaves. Many mangroves and saltbushes have these glands, which, seen under a microscope, look like little balloons. The salt is pushed into the balloons, which burst, releasing the salt back into the environment. Many of these strategies are similar to those found in metallophytes.

Atriplex prostrata

Metallophytes

Metallophytes are adapted to soils that contain large amounts of metal (often called heavy metals), which can be toxic to cellular processes. Like salts, they must be kept out of the plant using specialized strategies, or stored safely away from areas of active growth. Although the idea of industrial pollution may be the first thing that springs to mind when we talk about soils that contain metal, in fact many such soils are natural, including those formed from serpentine rocks. These rocks appear green because they contain high levels of nickel.

One way for metallophytes to store metals safely is by changing the metal into a safe, inactive molecule. They do this by binding other ionic molecules to the metal ions, preventing the metal ion from interfering with growth. Some metallophytes release these ionic molecules into the soil to inactivate the metal ions even before they are taken up, making the soil safer for other plants, too.

A few metallophytes are so good at storing and inactivating metals that these substances can build up in their tissues to a far greater level than in the surrounding soil. These are called hyperaccumulators. A tree native to serpentine soils of New Caledonia, *Pycnandra acuminata*, accumulates so much nickel that its sap appears blue-green.

Like halophytes, many metallophytes store accumulated metals in old leaves, which can be shed. Some, such as sea thrift (page 190), are both halophytes and metallophytes, and use their salt glands to get rid of metals as well.

Growing a mangrove in the backyard might be tricky, but many metallophytes – among them *Viola lutea*, *Minuartia verna*, *Odontarrhena serpyllifolium* (page 188), germander (page 187) and sea thrift – are rather lovely plants and worth considering for friendlier garden environments. The tough plants in this chapter will provide low-maintenance structure and colour that are suited to many gardens.

Lobularia maritima

Saxifraga species
Saxifrages

Dense mats of fine foliage, some silvery grey, others lime-green, spill out from between the rocks lining a sun-drenched path. Floating above the tight leafy cushions, pink flowers bob in the breeze. But don't be deceived by their delicate appearance; saxifrages are incredibly tough. Their Latin name even means 'stone breaker'.

Saxifraga is a large genus of plants that are native to wide swathes of the northern hemisphere and mountainous South America. This means that many can survive in very extreme conditions, from freezing alpine or Arctic temperatures to bone-dry soil and salty spray. Such species as *S. cespitosa* (shown opposite), *S. oppositifolia*, *S. bryoides* and *S. moschata* are native to rocky habitats, where the temperature plummets below -20°C (-4°F) in the winter. At the other end of the scale, summer temperatures can easily reach 30–35°C (86–95°F) in the rocky pockets (microsites; see pages 118–19) where the saxifrages grow. Even more challenging, these hot mountain microsites can be exposed to sudden icy storms in the middle of the summer, plunging temperatures below freezing. To survive such extremes, these incredible plants have evolved two types of adaptation: morphological (referring to growth shape) and biochemical (the chemicals they produce).

Many saxifrages have tiny or highly divided leaves, which lose less water from their surfaces simply because there is less surface from which to do so. Others have succulent leaves that store water. Many of these small or succulent leaves are also hairy, creating a humid layer close to the surface of the leaf and reducing the water that is lost to the atmosphere when the stomata are open to capture carbon dioxide.

Saxifrages also have taproots that store water. These roots are buried deep in the rocky crevices or sandy soils, not only providing access to water when rain is scarce, but also protecting the plant against freezing.

Plants that grow in stressful conditions also have biochemical adaptations. Saxifrages produce sugars and other molecules, such as amino acids, that act as antifreeze and slow the formation of ice in the tissues. Some of these molecules also increase the flexibility of the membranes, so that, as ice crystals form, the cells can swell with the ice, rather than bursting. These same chemicals also protect the membranes in drought conditions, when the cell membranes might shrink as water leaves the tissues.

These protective mechanisms render saxifrages incredibly tough. But despite this toughness, they are also amazingly beautiful, available in a wide diversity of flower colours and a range of leaf shapes and hues. They are great for any well-drained position, whether in a rockery, a sandy garden, a roof garden or a container.

Thymus serpyllum
Creeping thyme

A hot-pink carpet borders the gravel path, bright fingers reaching between the stepping stones. Creeping thyme is becoming a popular low-maintenance, pollinator-friendly alternative to grass for lawns. Producing a lovely scent when walked on, this attractive, dense ground cover needs very little attention once established because it is incredibly tolerant of both cold and dry conditions.

When water freezes, it can no longer be taken up by plants' roots, and this is why tolerance to cold and drought often go hand in hand in plants. A native of northern Europe and Siberia, creeping thyme is one of a range of thymes that can cope with extreme conditions. Their small leaves and commonly low-growing form protect them from drying winds, creating pockets of moist air among the dense leaves and allowing them to continue taking in carbon dioxide for growth without losing too much water.

Many thymes have a long taproot that allows access to water deep in the soil after the surface layer has dried out. They also produce protective molecules, such as 'antifreeze' (page 99) and osmolytes (page 80). Some of these are responsible for thyme's appealing scent and flavour, so dry conditions can enhance their culinary qualities.

Although most thymes produce low-growing carpets or small mounds, their foliage is very varied, ranging from deep green and blue-green to lime-green, and even variegated with yellow or white. Combined with flowers in hues from purple through pink to white, there's definitely thyme for everyone.

Asphodeline species
Asphodel/king's spear

Spears of lemon-yellow flowers rise above mounds of grass-like foliage. King's spear (*Asphodeline lutea*, shown here) is a striking plant that provides structure and colour in even the driest position.

One of the strategies that makes *Asphodeline* species drought-tolerant is the fleshy roots that can store water and resources, making these plants geophytes. When rain becomes infrequent, the leaves can use the resources that are stored in the roots, and because the roots penetrate deep into the ground, they can continue accessing water long after the surface layer has dried out. If conditions remain hot and dry for long periods, the shoots die back until things improve, when the shoots will resprout from sugars stored in the life-supporting roots.

Native to the margins of the Mediterranean and the Middle East, there are 17 species of *Asphodeline*, all with excellent drought tolerance. *A. liburnica* has yellow flowers set on fine stems, while others (such as *A. taurica* and *A. damascena*) have thicker stems carrying tall spears of white flowers.

This spear-like flower structure is reminiscent of another popular geophyte, the red-hot pokers (*Kniphofia* spp.), and of the flower spikes on *Xanthorrhoea* spp., the grasstrees among which I grew up in Australia. All three of these interesting and drought-tolerant plants are in the same family: Asphodelaceae. While *Asphodeline* can take five years to begin blooming, these tall structural spears, each proudly carrying rows of yellow or white flowers, are worth the wait.

Babiana species
Baboon

Tufts of vibrant green foliage appear to explode from the sandy soil, while purple flowers at ground level elbow their way between the leaves, apparently fighting for exposure to pollinators.

Babiana sambucina is a tough but beautiful African species, and the genus to which it belongs is locally called *Bobbejaantjie* (little baboon) after the chacma baboon, which eats the corms of these geophytes. When conditions are favourable, the plants absorb resources and store them in these corms; they then access them during times of drought. The leaves, which are covered in fine hairs, often die back before the onset of the hot summer, and on some species the stems remain entirely below ground at all times, providing additional protection from the summer heat.

The members of one interesting group of *Babiana* species, including *B. ringens* and *B. avicularis*, are bird-pollinated and grow stick-like perches for their feathery friends. It's very unusual for bird-pollinated flowers to be produced at ground level, where birds are at risk of predation. The vertical perches allow the birds to sidle down the stalk to collect the nectar (and with it pollen), while keeping a sharp eye out for danger.

There are many species and varieties of *Babiana* in a wide array of colours, including every shade of purple, red, pink, white and yellow, as well as multicoloured versions. Look out for *B. nervosa* (shown here; often called *B. stricta*), which is widely found in cultivation with many colourful flower options.

Teucrium species
Germanders

Bundles of silvery grey foliage are accented with small lilac-coloured flowers towards the end of the stems. Germanders are a lovely group of plants in the mint family (Lamiaceae).

With their silver foliage and deep roots, *Teucrium* species are good at tolerating dry conditions, by reflecting light and heat in order to keep cool, and accessing reserves of water deep in the soil. Aside from their tolerance to drought, many can also grow in soils that contain high levels of metals. Germanders are adapted to such metalliferous (metal-containing) soils and are able to keep the metals away from their actively growing tissues.

Germanders prefer chalky (calcareous) soils, and there are many different species, from tall shrubs (*T. fruticans*) to ground cover, such as *T. ackermannii* and *T. aroanium*. While those species with silvery reflective foliage are the most robust in hot, dry climates, others have bright green foliage, and these still have all the other adaptations that allow them to cope well with drought and metalliferous soils.

The flowers of germanders range in colour from shades of purple and pink to yellow (as in *T. scorodonia*) and white. These can be brightly presented, standing tall above the foliage, as in the dramatic *T. hircanicum* (shown here), or more subtle, peeping from among the foliage. An excellent choice for chalky soil, which can be a challenge for gardeners, germanders provide both structure and colour interest from their diverse growth forms, their silver or green foliage and their attractive flowers.

Odontarrhena species

Odontarrhena

A mass of yellow frothy flowers, below which fine grey-green foliage spills out of a crevice. Unusually, the surrounding rock is green, from the nickel it contains – and *Odontarrhena* is a plant that can tolerate high levels of metal in the soil.

All plants have in their cells specialized gates called transporters, which will move only one specific molecule. Nitrate, for example, is an important source of nitrogen for many plants, and it is moved into and around them by specialized nitrate transporters. Metallophytes also have specialized gates for moving metals. These transporters pump excess metal back into the soil, or move them around the plant so as to keep them away from the actively growing shoots or root tips. Many species of *Odontarrhena* can build up very high levels of nickel by safely stashing it in cell components where it will not damage the important processes needed for growth.

Odontarrhena used to be included within the genus *Alyssum*, and several species are still sold under that name, including *O. serpyllifolium* (shown here; often sold as *A. serpyllifolium*). It is one of many metallophytes that are also very tolerant of drought and salt, making it useful in many rocky gardens, whether or not the soil contains a lot of metal.

Lobularia maritima
Sweet alyssum

Frothy white flowers spill over the retaining wall, splashing towards the path like waves over a sea wall. A native of sand dunes around the Mediterranean, sweet alyssum is tolerant of both drought and salt, and has become widespread in gardens for its delightful flowers.

As with *Odontarrhena* (opposite), *Lobularia* used to be included in the genus *Alyssum*, hence the common name. This species is still often sold as *Alyssum maritima*.

Thanks to its origin by the sea, sweet alyssum can tolerate high levels of salt in the soil. It stores salts in parts of the cell where they are not able to interfere with the activities related to growth.

Salty soil also means that water from the plant wants to move out of the roots through osmosis. To turn the tide, *L. maritima* is one of those plants that accumulate ions, such as amino acids and sugars, in their cells. These are safe for cellular processes and help to draw water into the roots. This works when the soil is dry, too, so sweet alyssum is also drought-tolerant.

As an annual (or short-lived perennial), sweet alyssum will self-seed in most gardens. It is also available in a wide range of purple and pink forms, but for me the white is wonderfully reminiscent of the froth left on the beach as the waves recede.

Armeria maritima
Sea thrift

The sound of crashing waves surrounds an exposed rocky headland. At my feet, clouds of deep pink flowers float above a tight grassy lawn of bright green foliage.

Not surprisingly, given its often coastal rocky habitat, *Armeria maritima* has adapted to cold, salty and dry conditions. But it is also adapted to metalliferous soils, especially those that contain copper. To grow in these harsh and toxic conditions, sea thrift keeps its active growing shoots, roots and flowers safe from the salts and copper ions. Part of its strategy involves storing these toxic ions in old leaves, which it then drops.

Another way to get rid of copper and salt is through specialized salt glands on the leaf. Excess salt is stored in these glands, which burst when full, releasing the salt back into the environment. Sea thrift uses its salt glands to get rid of copper in the same way.

Not all sea thrifts come from salty or copper-filled soils, and some have adapted to easier conditions. This has led to localized populations that can be classified as the same species but which have developed different adaptations over generations. These local types are called ecotypes (or sometimes micropopulations) and are interesting examples of evolution in action.

Sea thrifts come in every shade of pink through to white, and are wonderful little plants for any well-drained position.

Atriplex species
Saltbushes/orache

Strikingly intense purple-red foliage catches my attention against the silvery gleam of the shrub beside it. Both are species of *Atriplex*. Thanks to their tolerance of drought and salt, such species as the purple *A. hortensis*, the silver-leaved *A. halimus* and the red-tipped *A. prostrata* (shown here) are increasing in popularity.

The saltbush genus includes species that are native to almost every part of the world, with a wide range of shapes, sizes and foliage colours. Many have specialized adaptations for living in salty habitats, such as salt lakes and salt marshes.

Several of the saltbushes have specialized salt glands on their leaves to collect excess salt. When the balloon-like gland is full, it bursts, releasing the salt back into the environment. If there is no rain to wash them away, the salt crystals can easily be seen on the surface of the leaf.

Many of the adaptations that make saltbushes well suited to salty soils also make them very tolerant of drought. With their lovely foliage, annual or perennial options, and a range of growth forms from ground cover to large shrubs, it's no wonder these tough plants are becoming popular. As a bonus, the leaves of some (including *A. hortensis*) even make a tasty alternative to spinach.

Yucca brevifolia
Joshua tree

A rocky vista stretches towards distant snow-capped mountains. Despite the hot, seemingly impossible conditions, the landscape is anything but desolate. Shrubs and grasses are accented by bizarre trees, some with single straight stems, some branching, each stem topped with a tuft of foliage. Their form is reminiscent of the Truffula trees from Dr Seuss's book *The Lorax* – but these are Joshua trees, *Yucca brevifolia*, and this is the Mojave Desert in the American Southwest.

Joshua trees have evolved a range of fascinating features that adapt them to hot, dry periods with intermittent rainfall. As do many yuccas, they have roots that contract, pulling the plant deeper into the ground. These contractile roots work by increasing their thickness, which in turn reduces their length. Being deeper in the ground protects against heat and helps to stabilize the tall stem in dry, shifting soil. Yuccas also produce extra (adventitious) roots from the base of the stem, so by pulling that base deeper into the soil, the plant has more stem surface below ground, from which new roots can emerge.

The succulent, water-storing leaves of Joshua trees are familiar to many gardeners, since yuccas have long been popular low-maintenance plants for the house and garden. Those leaves also have a very thick outer layer of wax to seal in moisture.

Such strategies are all about retaining moisture in this hot, dry environment, but Joshua trees also produce almost all their flowers on the tips of south-facing branches, which get the full hit of sun in the plant's northern-hemisphere native range. Even when flowers form on other branches, they bend around towards the south. To understand this, we first have to explore the special friendship yuccas have with moths.

Y. brevifolia has evolved to be pollinated exclusively by two species of yucca moth (*Tegeticula synthetica* and *T. antithetica*), which collect pollen by rolling it into their specialized tentacles. They then fly to the next flower, crawl inside and lay their eggs in the ovaries. Before leaving, they very deliberately rub some of the pollen they collected previously over the stigmas of the flowers. When the moth grubs hatch, they eat only some of the seeds. In this way the Joshua tree is entirely dependent on the moths for pollination, and the moths rely on the Joshua tree to feed their larvae.

South-facing branches warm up more than the shaded north-facing branches. As the sun sets, the warmer the flower, the longer it takes to lose that warmth, and this residual heat benefits the body temperature of the moths that are so crucial to the plant's' survival.

While not everyone can grow Joshua trees – they will not be happy in areas with long, dry summers, for example (see page 94) – there are many species of yucca, most from dry habitats, that will provide low-maintenance structural interest in a well-drained garden.

Resurrection plants
Miracle species that can dry out

A crumpled pile of dead leaves clings to the dry soil where once roots gave life to a plant. But is the plant as lifeless as it looks?

The members of one group of plants have become so good at surviving extremely dry conditions that they can appear dead until rain brings them back to health. Called resurrection plants after their ability to return from apparent death to life, these plants can put their cells to sleep, tucking them in with a blanket of protective sugars and amino acids. As the water slowly leaves, photosynthesis ceases and the foliage becomes dry and brown. Once in this dry state, they can sleep for months, even years, until rain awakes them. The crumpled leaves unfurl within minutes of rainfall, and if there is sufficient rain to soak into the dry soil and wake the roots, the cellular processes continue to revive, turning the leaves green and resuming growth.

Surely the most extreme of extremophiles, resurrection plants have evolved in a number of branches on the plant family tree, including mosses, ferns and fern allies, and flowering plants. A common example is *Selaginella lepidophylla*, a type of spikemoss in the family Selaginaceae, which curls up into a dry ball and unfurls its fronds once rewetted. It looks very similar to another common example, the rose of Jericho (*Anastatica hierochuntica*), a flowering plant in the mustard family (Brassicaceae).

These common examples are fun to watch on YouTube, but there are more than 300 known species of flowering resurrection plant, and many are truly beautiful. Let's begin with *Craterostigma plantagineum*, with its delicate purple and white flowers, each with two bright yellow guides signalling to pollinators. The flowers stand on short stems above the flat rosettes of leaves.

There are other resurrection plants that form small rosettes with lovely purple flowers, among them *Boea hygroscopica*, *Opithandra primuloides*, *Ramonda myconi* and *R. serbica*. *Haberlea rhodopensis* is another low-growing species that forms drifts in rocky spots, with tubular flowers that are purple on the outside and white inside. Others grow as dense mounds with starry flowers, among them *Borya inopinata* and *B. nitida*, and still others have tufts of grassy foliage and lily-like flowers, such as *Aylthonia blackii*.

Most of the species in the genus *Xerophyta* are resurrection plants. They range from small forms with orchid-like foliage and starry flowers (*X. humilis*) to the amazing *X. retinervis* (opposite), with its short, woody trunks sprouting long, grassy leaves and large, lily-like flowers.

With native options on almost every continent, it may come as a surprise that some of these resurrection plants are available in local nurseries. They prefer poor, well-drained soil, so if the garden soil is wet clay, a pot of free-draining sandy soil will suit them wonderfully, whether outside or in bright spots as a house plant. Perhaps the best thing about these amazing plants is that there is no need to worry about watering them while you are on holiday!

Punica granatum

Chapter 10

Edimentals: Food Plants with Adaptations

Introduction
Edimentals

Psophocarpus tetragonolobus

Along the path of stepping stones, barely visible among the creeping thyme, the brightly coloured blooms of *Brachyscome* and hardy geraniums crowd together where cyclamens and violets filled the winter months. Tall spears of *Asphodeline* and sharp, structural tufts of *Yucca* are offset by the soft greens, greys and purples of *Atriplex* and *Perilla* that grow tall against the low picket fence. Further down the garden, hawthorn and pomegranate give way to *Fuchsia boliviana*, and pineapple guava is planted between taller fruit trees, beneath which hostas thrive. Opening to a pond, the path winds between meadowsweet and loosestrife through sweet flag to water mint and watercress. Floating serenely on the surface, the wide, flat pads of waterlilies surround the delicate emerging flowers, while nearby a lotus stands regally, the first of its seed pods appearing among the elegant pale pink flowers. In the pond, spiked water milfoil wafts in the invisible currents made by the flick of a fish's tail. The hum of bees creates a relaxing backdrop to the calls of the many birds that flit among the trees and shrubs.

All this sounds idyllic – but it's also delicious! Every one of these species is edible in some way. Chosen as examples from the various chapters of this book, they are also great options for climate-resilient gardens. This final chapter widens that list of tough plants with those that have both ornamental and edible value. Commonly referred to as edimentals (edible ornamentals), they are by no means the only options, but rather are meant to whet the appetite for creating multifunctional gardens.

Caution!

Not all plants can be eaten, and some are extremely toxic. Similarly, even if one part of a plant is deliciously edible, another part of the same plant might be very toxic. This makes identification crucial. For local native plants, a good guide book can be incredibly helpful, and there are some excellent phone apps (my favourite is the Seek app by iNaturalist), but there is always some risk of misidentification. Combining these options with a local plant identification course (many nature reserves offer low-cost day courses or guided nature walks) is the best way to minimize the risk.

Bear in mind also that plants can take up whatever substances are in the ground, and these can then be in the parts of the plant that are eaten. Therefore, it's important to understand something of where the plant is growing. Some housing estates or nature reserves may be on sites that were once heavy industrial areas, while other locations could be naturally high in metals. Deep-rooted plants may take up metals from below the 'safe' layers of topsoil that are often added in new estates. If in doubt, use a soil-testing facility (available in most countries) to determine if there is anything to be concerned about in your garden.

While caution is needed, it is very rewarding to eat home-grown plants, and worth learning more about.

Allium carinatum

Edible plant parts

Now that you have learned about your local plants and satisfied yourself that the soil is safe, let's look at what part of the plants you can eat. Plants are made up of various organs, including roots, shoots, flowers and fruit, and all these can be edible.

Roots

Starting below ground, there is a surprising number of different structures that can make up a crop, from true roots to modified stems. Common examples are carrots (*Daucus carota*) and parsnips (*Pastinaca sativa*), which are swollen taproots from which the other roots sprout, and sweet potatoes (*Ipomoea batatas*), which are tubers. Both these types of underground storage organ are true roots.

Regular potatoes (*Solanum tuberosum*) are actually swollen modified (tuberous) stems. Similar examples include Jerusalem artichoke (*Helianthus tuberosus*) and daylilies (*Hemerocallis* spp. – also edible). Another type of modified stem is the rhizome, as seen on many members of the ginger family (Zingiberaceae; page 165) as well as the lotus (page 45) and species of *Canna*. Alliums – onions (page 208) – produce bulbs that are anatomically made up of layers of leaves, while taro (*Colocasia esculenta*) has a corm that looks very similar to a bulb. The difference is that corms are a swollen stem rather than the layers of swollen leaf bases that make up a bulb.

Shoots and flowers

In many plants, such as asparagus, the new shoots that emerge from the soil can be eaten. As they grow and expand, leaves emerge from the stems. In some plants, such as celery, both stems and leaves are edible, while in others, such as rhubarb, just the stems can be eaten. Many have edible leaves that can be used alongside lettuce in salads, such as shiso (page 209), saltbush (page 191), winged bean (page 206), amaranth (page 207), alliums and a wide range of flavoursome herbs, among them basil, mint and parsley.

Perhaps the least familiar edible plants in most diets are flowers. Courgettes (zucchini) and marrows have large flowers that can be eaten in salads or stuffed and cooked. In many plants, only the petals are edible and must be plucked from the rest of the flower structure. Aside from the examples in this chapter, edible flowers include *Dianthus* spp. (pinks), chrysanthemums, daylilies, elderflower (*Sambucus nigra*), hibiscus, hollyhocks (*Alcea* spp.), pansies and violets (*Viola* spp.).

Amaranthus tricolor

Seeds, nuts and fruit

Finally, after the remaining flowers are pollinated, fruit and seeds form. The seeds of some plants, such as amaranth, can be ground in the manner of wheat, rice and maize to make flour. Others, including cardamom, cumin, dill, fennel, mustard, nutmeg and Szechuan pepper, are used as spices in a wide range of cuisines across the world. Nut-like seeds, such as macadamias, chestnuts, pecans, Brazil nuts and almonds, are common in most supermarkets, but other edible types include beech (*Fagus sylvatica*) seeds, *Elaeocarpus* nuts, ginkgo, baru (*Dipteryx alata*), butternut (*Juglans cinerea*), *Canarium* nuts, lotus seeds and saba (*Pachira aquatica*), to name just a few.

Other plants produce their seeds in juicy fruit that will be dispersed by animals. Forming mostly from the ovary of the flower, fruit can be simple, like a grape or redcurrant, or aggregate, as in the raspberry, the flower of which contains bunches of ovaries that grow to form a single fruit. By contrast, pineapples, figs and mulberries are made from a whole bunch of flowers (called an inflorescence), and are called multiple fruit. Finally, apples and pears include flower parts other than the ovary, and are called accessory fruit (think of the other flower parts as accessories).

Feijoa sellowiana

Edimentals

In addition to the mainstream fruit and vegetables that are available in the supermarket, a wide range of plants are edible and many are also very attractive. Combining trees, shrubs, perennials and annuals can bring year-round interest to both the garden and the plate. There are also plenty of options that can be grown in pots or up a trellis, making them useful for small gardens and even balconies. This chapter introduces a few that will prove resilient in different climates.

Phyteuma spicatum and *Campanula rapunculus*

Spiked rampion/Rapunzel

Tall, creamy flower spikes, their individual flowers opening from the base of the flower head, the top ones still closed like many curved fingers reaching for the sky. At first glance, spiked rampion could not look more different from the blue, bell-like flowers of the rampion bellflower (*Campanula rapunculus*; opposite, left), yet both are in the family Campanulaceae and they share an interesting history.

There once lived a man and his wife, who had long wished in vain for a child. One day the wife saw the finest rampion growing in the witch's garden behind their house, and she longed for it. Rather than have his wife pine away, the man climbed over the wall, plucked a handful of the leaves and brought them to her. She liked them so much that she needed more, so the next evening he went again into the forbidden garden – but this time, as he was leaving, clutching a handful of the delicious leaves, the witch appeared. The man explained his wife's situation and the witch offered them all the rampion they liked, on one condition: that they give her the child they would soon bring into the world. The husband agreed, and when the child came, the witch whisked away the little girl, naming her after the rampion – Rapunzel.

There has been much discussion about which species is described in the Brothers Grimm's story of Rapunzel, but the two most likely are *Phyteuma spicatum* (opposite, right) and *Campanula rapunculus*. Both are attractive, edible plants that can be used in similar ways. The leaves can be eaten raw in salads, decorated with the flowers (which are also edible), or cooked like spinach, and are very high in vitamin C. The unopened flowers of *P. spicatum* can also be steamed, while the young shoots of *C. rapunculus* can be eaten in the same way as asparagus.

Below ground are the tuberous, nutty-flavoured roots of both species, which can be boiled and used in salads or roasted with other root vegetables. These delicious tubers are an adaptation that helps these species to thrive in poor conditions. Storing nutrients and sugars, they survive periods of drought or stay warm in the soil during the winter, ready to sprout those tasty new shoots in the spring. Happy on any well-drained soil, these rampions will require little attention to provide their attractive crop.

These two are not the only edible members of the family. Many have tasty leaves, including *C. versicolor*, *C. portenschlagiana* and *C. medium*. Others have sweet edible roots, including *Adenophora latifolia*, *A. liliifolia* and *Codonopsis lanceolata*. The last of these three is a climber with green bell-shaped flowers that are deep purple inside, and is popular in Korean food as *deodeok*.

Many of these species are familiar garden ornamentals, but it's worth remembering the story of Rapunzel and using these beautiful plants as versatile edibles.

Persicaria odorata
Vietnamese mint

Brown arrows over vibrant green point towards the tips of long, narrow leaves in this attractive herb with dainty pink flowers.

A native of tropical and subtropical climates, *Persicaria odorata* thrives in warm, damp conditions but can grow in drier soil in cool climates. Where the days become very hot, it prefers growing in shade, and it can also thrive underwater, as long as some of its leaves peek out above the surface.

One of the traits that make this plant so adaptable is its ability to grow roots along its stems. To see them (and to propagate the plant), pop a piece of stem in a glass of water; new white roots will emerge within a few days. This adaptability comes with a caution, however. Many people report that this delicious herb can take over the garden, so restricting it to a pot may be the wise option.

Unsurprisingly, given its common name, this plant is used regularly in Vietnamese dishes. The leaves have a fresh flavour, somewhat like that of coriander (cilantro), although without the soapy flavour that some people experience with the latter. Delicious in salads and freshly shredded into bowls of the South Asian noodle soup known as laksa, it's also added to Vietnamese fish stews.

The closely related species *P. minor* can replace Vietnamese mint in cool climates, while alpine bistort (*Bistorta vivipara*), in the same family (Polygonaceae), is very tolerant of cold, and its seeds and roots are delicious. From the Vietnamese mint of the humid tropics to *Bistorta vivipara* of alpine gardens, there is a tasty member of this family for any location.

Lathyrus tuberosus
Earthnut pea

Intensely pink, sweetly scented pea flowers are held on stalks away from lush foliage that climbs with tightly curled tendrils. *Lathyrus tuberosus* is a delightful member of the sweet peas, and to unearth its food value we must dig it up.

While members of the genus *Lathyrus* look similar to the related garden peas (*Pisum sativum*) of which we eat the juicy seeds or the new sprouts as microgreens, sweet peas have toxic seeds. The earthnut pea, however, produces small underground tubers that take a few years to develop but which have a lovely nutty flavour when roasted or boiled.

The tubers also help the earthnut pea to survive periods of dry weather, storing resources safely below the ground. This legume is happy on poor soils, and its roots produce nodules that house helpful bacteria. These bacteria turn nitrogen in the air into fertilizer for the plant, and in return the plant feeds the bacteria with sugars to help them grow.

Native across Europe, the Middle East and northern parts of China, the earthnut pea is adaptable to a wide range of soils and conditions. But this very adaptability brings the risk that it will become invasive, and indeed it is considered a noxious weed in parts of the US, so, as always, check the restrictions in your area before planting it in your garden.

Requiring very little attention and having both delicious tubers and beautiful, highly scented flowers that attract pollinators, it's easy to understand why earthnut pea is regaining popularity in allotments.

Psophocarpus tetragonolobus
Winged bean

Cascades of deep green foliage studded with pale purple pea flowers spill over the support, while long four-cornered pods hang down ready for harvest. The winged bean can certainly hold its own among ornamentals, but in fact every part of this attractive plant is edible.

The leaves can be used in a similar way to spinach, the young pods eaten raw or cooked like runner beans, and the flowers used in salads. Note that the seeds contain toxins, which are broken down by heating, so must be eaten roasted, fried, boiled or steamed. Below the ground, the winged bean grows nutty-tasting tubers that can be eaten boiled, baked, fried or steamed, in the same way as potatoes. In common with many seeds and legumes, the beans of winged bean are high in protein, but the tubers, too, are among the most protein-rich of any root vegetable.

This native of eastern tropical Africa has long been grown widely across Asia and is now gaining a following in other warm areas thanks to its versatility. Not only can every part be eaten, but also it is incredibly tolerant of a wide range of conditions. Despite being happiest in warm, humid climates with tropical rainfall and well-drained soil, it is very tolerant of dry conditions or heavy soil, so long as it does not become waterlogged.

With its attractive purple flowers, interesting seed pods, lovely green foliage and habit that allows it to be grown vertically, taking up minimal space in the garden, the winged bean certainly puts the 'edible' in 'edimental'.

Amaranthus tricolor
Amaranth/Chinese spinach

A tall, lush shrub, its maroon-hearted leaves edged with vibrant green, is draped with pink, feather boa-like flowers. *Amaranthus tricolor* is stunning in every way.

But don't be fooled by its fancy dress – amaranth is a surprisingly resilient species. Native to the tropics, it also grows happily in mild temperate regions. Where winter temperatures drop below freezing, it can be grown as an annual or brought indoors during the cold months.

Amaranth is very drought-tolerant, thanks to both its deep roots and its use of C4 photosynthesis (see page 28). This method conserves water by allowing sugar production to continue even while the stomata are closed during the heat of the day.

Some amaranth cultivars have multicoloured leaves splashed with green, red and yellow. Regardless of the colour intensity, however, they can be eaten raw in salads, stir-fried or used in soups. The seeds can be ground into flour, popped like popcorn or flaked like oatmeal. In Mexico, popped amaranth seed is used to make a treat called *alegria*, which – appropriately for this showy plant – means 'happiness' in Spanish.

Allium carinatum
Keeled garlic

Flowers like fireworks of purple and white explode from slender green stems. With its attractive onion-like flowers, keeled garlic would be at home in any cottage garden.

Below ground, small edible bulbs form, storing resources during cold or dry conditions. This fully hardy plant will grow on a range of soil types, although it prefers not to have soggy roots.

The new shoots appear from winter onwards and can be eaten in the same way as chives when other greens are scarce. In fact, every part of this plant can be eaten. The bulbs can be used like onions, the flowers can be eaten in salads, and the leaves can be cooked or enjoyed raw. The flavour, which falls somewhere between that of onions and garlic, is down to the type of sulphur compounds in the plant. This is true of all onions, and if the soil is deficient in sulphur, the bulbs will be tasteless and bland.

The many types of commercial onion are all variants of a single species: *Allium cepa*. From this come brown, white, yellow and red onions, shallots and spring onions (scallions). Other edimental alliums include *A. roseum*, with its bundles of pale pink edible flowers, the white-blooming *A. zebdanense* and yellow *A. moly*. While many edible flowers are more impressive for their appearance than for their taste, those of *Allium* are anything but bland, bringing a chivey, oniony kick to any salad.

Perilla frutescens
Shiso/beefsteak plant

A dense shrub with frilly-edged purple leaves, *Perilla frutescens* is a tall perennial in the mint family (Lamiaceae). With an unusual flavour that is often described as being somewhere between mint and basil, the leaves are widely used as a herb, eaten fresh in salads, or cooked in stews, soups and stir-fries.

P. frutescens encompasses a range of varieties with leaves that are frilled or flat with a jagged, toothed edge, in shades of green through to purple and vibrant red, from which the common name 'beefsteak plant' originates. A thriving perennial in humid warmth, shiso is happy in just about any soil type, and can be grown as an annual in cool climates.

Native across Asia, shiso is widely used in Asian cuisines. Most of the plant can be eaten, including the flowers, which can be cooked, and the seeds, which can be toasted and used as a spice. Depending on the cuisine, the leaves of various colours are used for different purposes. The red leaves are often used to bring colour to dishes, and the green are used as a garnish.

The plant also produces flower spikes in shades of pink and white that stand above the foliage. Popular with pollinators, shiso will self-seed around the garden. In the right conditions, it can become weedy, so in some regions care must be taken that it does not become invasive. But if local regulations allow, this attractive edible plant with its vibrant foliage and pollinator-friendly flowers is well worth growing and eating.

Abelmoschus esculentus
Okra/ladies' fingers

Pale yellow hibiscus-like flowers shine from green foliage, and fruit like green fingers points up from between the leaves. Often called ladies' fingers after the shape of these pods, okra is a widespread crop that is gaining mainstream attention in the West.

Having originated in India and neighbouring countries, okra is now widespread across the tropics and subtropics, including throughout Africa. From the 16th century onwards it was brought to the southern states of the US by enslaved Africans, and there its seeds were used to make a caffeine-free coffee substitute during the American Civil War (1861–5).

While those of us who prefer our coffee caffeinated may not be ready for this substitute, there is another preparation native to Louisiana that uses the green fruit of okra: the meat or fish stew known as gumbo. The notorious slime from the okra pods is used as a thickener in this dish, the okra being cooked first and the other ingredients added once the consistency is right. To reduce the slimy effect in other dishes, fry the okra with acidic ingredients, such as tomatoes. The leaves can also be added to salads.

In common with many species in the mallow family (Malvaceae), okra is very drought-tolerant and can grow on almost any soil, although it won't tolerate freezing, so in temperate areas it's often grown as an annual. There are also red-fruited varieties, such as 'Red Burgundy' and 'Little Lucy', the latter of which also has dark green leaves mottled with deepest red. So even if the slimy pods are not to everyone's taste, the edible leaves and attractive large flowers make okra a valuable edimental.

Feijoa sellowiana
Pineapple guava/feijoa

Bright red paintbrushes tipped with yellow pollen adorn the dense canopy that shimmers in the breeze, as the silvery undersides of the leaves flicker against the dark green of the leaf tops. Pineapple guava is an attractive, small multistemmed tree that is native to eastern South America.

Despite its tropical appearance, feijoa actually requires some winter chilling if it is to fruit, so it won't fruit in the tropics. Tolerant of frost and happy in full sun or partial shade, it will also cope with periods of drought once it is established, and it's happy in most soils as long as its roots don't become soggy.

The beautiful flowers, which attract birds, are edible and sweet-smelling, and can be added to salads, iced tea and jelly (jello). But the main event is the fruit. Similar in size to a kiwi and also green, they have been described as a cross between pineapple, banana, guava and kiwi in flavour, but perhaps this unique taste should simply be described as like a pineapple guava! The fruit is ripe when it falls off the tree, and can be cut open and the flesh scooped out with a spoon. Eat it then and there, or add it to ice cream, smoothies or juices.

As a small, slow-growing tree, feijoa is great for small gardens and can also be grown in a pot on a balcony. Regardless of garden size, these beautiful trees can be grown in a surprisingly wide range of climates, bringing a tropical appearance and bearing their tasty rewards for us to enjoy.

Punica granatum
Pomegranate

Shimmering red jewels spill from swollen fruit. A study in scarlet, almost every part of the pomegranate's flowers are bright red, interrupted only by the sunny yellow pollen-covered anthers. Once pollinated, the ovaries swell to produce the red fruit (technically a berry) that contains the deliciously juicy seeds.

Native to the Middle East, this small, often multistemmed tree is incredibly resilient. Its long roots seek water deep in the ground, and it is happy in a wide range of soils, including clay, loamy, pebbly, sandy and even salty. The thick, glossy leaves reflect excess light and heat, and provide dense shade against the relentless heat of summer in its native range.

The pomegranate's resilience doesn't stop there. The winter temperatures of the semi-arid regions from which it originates can plummet well below freezing. To cope with such icy conditions, the tree enters a state of dormancy by gradual steps triggered by a combination of shortening days and decreasing temperature. By midwinter it is fully dormant and has produced molecules, such as sugars and amino acids, that act as antifreeze, protecting the cell membranes from damage.

Freezing conditions also turn water to ice, making it unavailable to plants. The sugars and amino acids help to manage this winter drought as they would a summer shortage. As the days begin to lengthen in early spring, growth resumes and the scarlet flowers open from late spring, continuing into summer.

There are many varieties of pomegranate, each with its own qualities, including fruit in different colours from orange through shades of red to purple. Some are better suited than others to particular environments, so choose one for your local climate. There are also varieties with even more interesting flowers: edged in white, or frilly-skirted with two layers of petals. Many of those bred for ornamental flowers will not produce fruit, however, so if you want an edimental, stick to the still-beautiful fruiting options.

All pomegranates are relatively small trees, but there are properly dwarf varieties that will grow happily in pots. In locations with cool or short summers, they can be moved to the conservatory or greenhouse to extend the period of warmth and ripen the fruit.

Health-giving, delicious and able to grow in challenging environments, it's not surprising that the pomegranate appears throughout recorded history. In ancient Egypt it was a symbol of prosperity, fertility and eternal life, and in ancient Greece pomegranates were given to new homeowners to bring fertility and good luck – a custom that persists today. Science is unravelling the health benefits of this juicy, antioxidant-filled fruit.

Valued in many cuisines and suited to lamb and chicken dishes, soups and tagines, the pomegranate is for me the taste of summer, whether sprinkled on a salad or drunk as juice. Whatever your taste, pomegranates make an attractive and versatile edimental for some of the toughest conditions.

Growing food
The long-term approach

In a jungle of colour, tall, intensely purple tree collard grows among Korean celery with white umbel flowers the size of dinner plates. Perennial onions flower away, while walking onions produce aerial bulbs that loop to the ground like a caterpillar in search of the next leaf. Over the wooden fence rambles a beautiful, lush-leaved, dusky maroon-and-pink-flowering hopniss, also called groundnut for its protein-rich tubers.

Surrounded by the hum of bees and birdsong, it's easy to forget that this beautiful scene is entirely edible. Mandy Barber of Incredible Vegetables in Devon, southwestern England, tells me that perennial food gardens are all about shared space: shared with other plants (even 'weeds', which might be food for insects, for example); with birds, insects and other critters that make the space their home; and with humans, for food and beauty.

Conventionally grown crops are largely annuals, which are sown into ploughed fields, grown for a season, harvested, then reploughed ready for the next year's crop. This interrupts the processes of soil biology, and although it works for large-scale harvesting, it restricts fields to a single crop.

Plenty of edibles are truly beautiful and happy growing among other plants, so allotments and gardens can be diverse and productive. Think about a woodland. The plants don't all grow in rows of a single species, with nothing else around them. Instead, they weave together, each growing in the patch it likes the best. Edibles are exactly the same. An interwoven, multilayered strategy will yield much more food than a monoculture.

One of Barber's tips is to be wise about space. The perennial collard, for example, grows vertically, and its leaves can be harvested all along the stem, producing far more than its annual relatives. Also look at what is cheap in the supermarket – and don't grow that. Focus instead on more expensive products that will save you the most money. Potatoes might be easy to grow, but they take up a lot of space and are usually cheap to buy in the shops.

While perennials are more productive than annuals once they are up and growing, it can require patience to get them started, so it's useful to have a few quick fixes, especially those that are ready early in the season, such as radishes. These grow fast and can be harvested quickly, minimizing the use of space over time and giving that harvesting high that inspires us to keep going.

Many perennial vegetable seeds need a period of cold before they will germinate, so they can be sown in late autumn and left outside or popped in the fridge for a few weeks. Most will then germinate over the following weeks, although some can take longer – so be patient.

Some of Barber's favourite plants are perennial brassicas, such as the beautiful purple tree collard (*Brassica oleracea* var. *acephala*) and perennial kale (*B. o.* var. *ramosa*), which produce prolifically and are easy to propagate from cuttings. The walking onion (*Allium × proliferum*) and everlasting onion (*A. cepa* 'Perutile'), meanwhile, grow in clumps that can be repeatedly divided, leaving some to keep growing while the others are eaten.

For those starting from scratch, Barber advises allowing so-called weeds to share the space. Keep the soil structure intact by following a no dig-method, adding organic matter on the surface and letting the worms do the work of incorporating it.

As a rallying call to us all, the edible perennial world needs our imagination. What grows well with what is still relatively unexplored. Whether for designing attractive spaces or trying new culinary delights, there's a great deal for us still to find out.

Glossary

Adventitious root A root that forms from anything other than another root.

Annual A plant that lives its whole life within one year.

Biennial A plant that lives its life over two years, usually growing in year one and flowering in year two.

Bract A modified leaf at the base of a flower or flower stem (e.g. the red leaves of poinsettia or the bright leaves of Japanese dogwood, *Cornus kousa*).

Chlorophyll The green pigment that captures light for **photosynthesis**.

Cross-pollination When pollen from one plant pollinates a different plant.

Cultivar A plant variety that has been produced in cultivation by selective breeding.

Cuticle A waxy layer secreted from a leaf's outer cell layer.

Deciduous A plant that drops its leaves in the autumn/fall.

Dicot A flowering plant of which the embryo has two seed leaves.

Epidermis The outermost layer of cells of a plant organ (including leaves, stems and roots).

Genus (plural **genera**) Based on Linnaean classification, a group containing related plants.

Geophyte A plant with growing points that rest below the ground when conditions are difficult (includes those that produce **rhizomes** and bulbs).

Herbaceous A plant with soft shoots and non-woody stems; usually an **annual** or a **perennial** that dies back to the ground in the winter.

Heterophylly Having different types of leaf on the same plant (e.g. holly, which has sharply pointed young leaves and smooth-edged mature leaves).

Internode The distance between **nodes** on a stem.

Legume A plant of the family Fabaceae that makes nodules on roots to fix nitrogen from the air.

Monocot A flowering plant with a single seed leaf on the embryo, and typically parallel veins (e.g. grasses).

Node A point on the stem where a leaf is attached.

Osmosis The movement of water through a membrane from an area of high concentration of salts to an area of low concentration, typically to balance out the salts.

Perennial A plant that lives for more than two years.

Photosynthesis The process of harvesting light and carbon dioxide and converting it into sugars and oxygen.

Rhizome An underground stem, often used to store resources.

Spore A single reproductive cell. In mosses and fungi spores germinate to make the next generation of plants; in conifers and flowering plants spores are involved in pollination.

Stolon A stem that grows above or on the ground and can create new plants (e.g. in the strawberry).

Stomate or **stoma** (plural **stomata**) A pore in the leaf that can open and close to control **transpiration**.

Substrate The substance a plant is growing in or climbing on (e.g. soil, potting mix, wall, tree trunk).

Transpiration The movement of water from inside the leaf, through the **stomata**, into the atmosphere.

Bibliography

Introduction

Ali, N S, and C Trivedi, 'Botanic Gardens and Climate Change', *Biodiversity & Conservation* 20 (2011), 295–307

Bak, J, and M Barjenbruch, 'Benefits, Inconveniences, and Facilities of the Application of Rain Gardens in Urban Spaces from the Perspective of Climate Change: A Review', *Water* 14/7 (2022), 1153

'Causes and Effects of Climate Change', United Nations, www.un.org/en/climatechange/science/causes-effects-climate-change (accessed Jul 2024)

Dullinger, I, et al., 'Climate Change Will Increase the Naturalization Risk from Garden Plants in Europe', *Global Ecology & Biogeography* 26/1 (Jan 2017), 43–53

Haeuser, E, et al., 'European Ornamental Garden Flora as an Invasion Debt under Climate Change', *Journal of Applied Ecology* 55/5 (Sep 2018), 2386–95

Primack, R B, et al., 'The Growing and Vital Role of Botanical Gardens in Climate Change Research', *New Phytologist* 231/3 (Aug 2021), 917–32

Van der Veken, S, et al., 'Garden Plants Get a Head Start on Climate Change', *Frontiers in Ecology and the Environment* 6/4 (May 2008), 212–16

Chapter 1

Kettley, S, and B Evans, 'The Palm that Flowers Underground', 28 June 2023, www.kew.org/read-and-watch/the-palm-that-flowers-underground

Santi, C, et al., 'Biological Nitrogen Fixation in Non-legume Plants', *Annals of Botany* 111/5 (May 2013), 743–67

Chapter 2

'5 Ways to Create a Wildlife Friendly Garden', Nature Conservancy Australia, 6 August 2020, www.natureaustralia.org.au/get-involved/take-action/wildlife-friendly-garden

Busch, J, 'Characteristic Values of Key Ecophysiological Parameters in the Genus *Carex*', *Flora* 196/6 (2001), 405–30

Colmer, T D, and O Pedersen, 'Underwater Photosynthesis and Respiration in Leaves of Submerged Wetland Plants', *New Phytologist* 177/4 (2008), 918–26

Erbil, H Y, et al., 'Transformation of a Simple Plastic into a Super-hydrophobic Surface', *Science* 299/5611 (Feb 2003), 1377

Gibbs, J, et al., 'Regulation of Alcoholic Fermentation in Coleoptiles of Two Rice Cultivars Differing in Tolerance to Anoxia', *Journal of Experimental Botany* 51/345 (Apr 2000), 785–96

Grant, N M, et al., 'Distribution of Thermogenic Activity in Floral Tissues of *Nelumbo nucifera*', *Functional Plant Biology* 37/11 (Oct 2010), 1085–95

'How to Provide Water for Wildlife', The Wildlife Trusts, www.wildlifetrusts.org/actions/how-provide-water-wildlife (accessed Jul 2024)

Lenssen, J P M, et al., 'Vegetative Reproduction by Species with Different Adaptations to Shallow-flooded Habitats', *New Phytologist* 145/1 (Jan 2000), 61–70

Li, H-T, et al., 'Plastid Phylogenomic Insights into Relationships of All Flowering Plant Families', *BMC Biology* 19 (2021), article no. 232

Li, J-K, and S-Q Huang, 'Flower Thermoregulation Facilitates Fertilization in Asian Sacred Lotus', *Annals of Botany* 103/7 (May 2009), 1159–63

Maberly, S C, and T V Madsen, 'Freshwater Angiosperm Carbon Concentrating Mechanisms', *Functional Plant Biology* 29/3 (Apr 2002), 393–405

Madsen, T V, 'Growth and Photosynthetic Acclimation by *Ranunculus aquatilis* L. in Response to Inorganic Carbon Availability', *New Phytologist* 125/4 (Dec 1993), 707–15

——, and S C Maberly, 'High Internal Resistance to CO^2 Uptake by Submerged Macrophytes that Use HCO^{3-}', *Photosynthesis Research* 77 (2003), 183–90

Seymour, R S, and P Schultze-Motel, 'Physiological Temperature Regulation by Flowers of the Sacred Lotus', *Philosophical Transactions: Biological Sciences* 353/1371 (Jun 1998), 935–43

Van Veen, H, and R Sasidharan, 'Shape Shifting by Amphibious Plants in Dynamic Hydrological Niches', *New Phytologist* 229/1 (Jan 2021), 79–84

Visser, E J W, et al., 'Flooding Tolerance of Carex Species in Relation to Field Distribution and Aerenchyma Formation', *New Phytologist* 148/1 (Oct 2000), 93–103

'Water', National Wildlife Federation, www.nwf.org/garden-for-wildlife/water (accessed Jul 2024)

Yu, X, et al., 'Transcriptome Analysis and Morphological Changes in Response to Waterlogging in *Iris pseudacorus*', *Phyton* 91/10 (2022), 2135–62

Chapter 3

Braendle, R, and R M M Crawford, 'Plants as Amphibians', *Perspectives in Plant Ecology, Evolution and Systematics* 2/1 (1999), 56–78

'How to Grow Carnivorous Plants', RHS Gardening, www.rhs.org.uk/plants/types/carnivorous/growing-guide (accessed Jul 2024)

Kratsch, H A, and William R. Graves, 'Location and Anatomy of Nodules on *Alnus maritima* Subjected to Flooding', *Journal of the American Society for Horticultural Science* 129/6 (2004), 775–80

Nakamura, M, and K Noguchi, 'Tolerant Mechanisms to O_2 Deficiency under Submergence Conditions in Plants', *Journal of Plant Research* 133 (2020), 343–71

Nisar, S, et al., 'Salicylic Acid Retards Senescence and Makes Flowers Last Longer in *Nicotiana plumbaginifolia* (Viv)', *Plant Physiology Reports* 26 (2021), 128–36

Pai, A, and B C McCarthy, 'Variation in Shoot Density and Rhizome Biomass of *Acorus calamus* L. with Respect to Environment', *Castanea*, 70/4 (Dec 2005), 263–75

Romanello, G A, et al., 'Morphological Adjustments Promote Drought Avoidance in the Wetland Plant *Acorus americanus*', *Aquatic Botany* 89/4 (Nov 2008), 390–6

Seymour, R S, et al., 'Effects of Floral Thermogenesis on Pollen Function in Asian Skunk Cabbage *Symplocarpus renifolius*', *Biology Letters* 5/4 (Aug 2009), 5568–70

Smirnoff, N, and R M M Crawford, 'Variation in the Structure and Response to Flooding of Root Aerenchyma in Some Wetland Plants', *Annals of Botany* 51/2 (Feb 1983), 237–49

Smith, A M, et al., 'Alcohol Dehydrogenase Activity in the Roots of Marsh Plants in Naturally Waterlogged Soils', *Planta* 168/1 (May 1986), 130–8

Stevens, K J, et al., 'The Aerenchymatous Phellem of *Lythrum salicaria* (L.)', *Annals of Botany* 89/5 (2002), 621–5

Tanaka, K, et al., 'Morphological and Anatomical Changes of *Melaleuca cajuputi* under Submergence', *Trees* 25 (2011), 695–704

Thorogood, C J, et al., 'Convergent and Divergent Evolution in Carnivorous Pitcher Plant Traps', *New Phytologist* 217/3 (Feb 2018), 1035–41

Van Veen, H, et al., 'Two Rumex Species from Contrasting Hydrological Niches Regulate Flooding Tolerance through Distinct Mechanisms', *Plant Cell* 25/11 (Nov 2013), 4691–707

Wenlin, W, et al., 'Spatio-temporal Patterns in Rhizosphere Oxygen Profiles in the Emergent Plant Species *Acorus calamus*', *PLOS ONE* 9/5 (2014), e98457

Woolfson, K N, et al., 'Suberin Biosynthesis, Assembly, and Regulation', *Plants* 11/4 (Feb 2022), 555

Chapter 4

Bermúdez, R, and R Retuerto, 'A Sunny Day at the Beach: Ecophysiological Assessment of the Photosynthetic Adaptability of Coastal Dune Perennial Herbs by Chlorophyll Fluorescence Parameters', *Photosynthetica* 52/3 (2014), 444–55

Berry, Z C, et al., 'Foliar Water Uptake', *Plant, Cell & Environment* 42/2 (Feb 2019), 410–23

Blanusa, T, et al., 'Alternatives to *Sedum* on Green Roofs', *Building & Environment* 59 (Jan 2013), 99–106

Bordbar, F, et al., 'Leaf Anatomical Investigations in

Acantholimon (Plumbaginaceae)', *Brazilian Journal of Botany* 45 (2022), 729–41
Caser, M, et al., 'Ecophysiological and Phytochemical Responses of *Salvia sinaloensis* Fern. to Drought Stress', *Plant Growth Regulation* 84 (2018), 383–94
Dörken, V M, 'Leaf-morphology and Leaf-anatomy in *Ephedra altissima* Desf. (Ephedraceae, Gnetales) and their Evolutionary Relevance', *Feddes Repertorium* 123/4 (Feb 2014), 243–55
Ferrante, A, et al., 'Physiological and Morpho-anatomical Traits Used as Markers for the Selection of Drought Tolerance of Ornamental Plants', *Acta Horticulturae* 1331 (2021), 253–60
Funk, J L, and V A Zachary, 'Physiological Responses to Short-term Water and Light Stress in Native and Invasive Plant Species in Southern California', *Biological Invasions* 12 (2010), 1685–94
Gorai, M, et al., 'Rapid Recovery of Photosynthesis and Water Relations Following Soil Drying and Re-watering Is Related to the Adaptation of Desert Shrub *Ephedra alata* subsp. *alenda* (Ephedraceae) to Arid Environments', *Environmental & Experimental Botany* 109 (Jan 2015), 113–21
Herrera, A, 'Crassulacean Acid Metabolism-cycling in *Euphorbia milii*', *AoB Plants* 5 (2013), plt014
Isermann, M, and P Rooney, 'Biological Flora of the British Isles: *Eryngium maritimum*', *Journal of Ecology* 102/3 (May 2014), 789–821
Kayabaş, A, 'SEM Observations on the Vegetative Plant Parts of *Acantholimon riyatguelii* Yıldırım (Plumbaginaceae), a Local Endemic Restricted to Gypsum Habitats', *Journal of Structural Biology* 214/1 (Mar 2022), 107830
Luo, D-Y, et al., 'Repellency and Insecticidal Activity of Seven Mugwort (*Artemisia argyi*) Essential Oils against the Malaria Vector *Anopheles sinensis*', *Scientific Reports* 12 (2022), article no. 5337
Neumann, R B, and Z G Cardon, 'The Magnitude of Hydraulic Redistribution by Plant Roots', *New Phytologist* 194/2 (Apr 2012), 337–52
Pan, Z, et al., 'The Upside-down Water Collection System of *Syntrichia caninervis*', *Nature Plants* 2 (2016), 16076
Šantrůček, J, 'The Why and How of Sunken Stomata', *Annals of Botany* 130/3 (Sep 2022), 285–300
Schwinning, S, et al., 'Deuterium Enriched Irrigation Indicates Different Forms of Rain Use in Shrub/Grass Species of the Colorado Plateau', *Oecologia* 130 (2002), 345–55
Shane, M W, and H Lambers, 'Systemic Suppression of Cluster-root Formation and Net P-uptake Rates in *Grevillea crithmifolia* at Elevated P Supply', *Journal of Experimental Botany* 57/2 (Jan 2006), 413–23
Singh, M, et al., 'Roles of Osmoprotectants in Improving Salinity and Drought Tolerance in Plants: A Review', *Reviews in Environmental Science & Bio/Technology* 14 (2015), 407–26
Ustin, S L, and S Jacquemoud, 'How the Optical Properties of Leaves Modify the Absorption and Scattering of Energy and Enhance Leaf Functionality', in J Cavender-Bares et al. (eds), *Remote Sensing of Plant Biodiversity* (Cham, 2020)
Vaz Monteiro, M, et al., 'Relative Importance of Transpiration Rate and Leaf Morphological Traits for the Regulation of Leaf Temperature', *Australian Journal of Botany* 64/1 (2016), 32–44

Chapter 5

Azocar, A, et al., 'Freezing Tolerance in *Draba chionophila*, a "Miniature" Caulescent Rosette Species', *Oecologia* 75/1 (Feb 1988), 156–60
Herrera, C M, and M I Prozo, 'Nectar Yeasts Warm the Flowers of a Winter-blooming Plant', *Proceedings of the Royal Society B* 277/1689 (Jun 2010), 1827–34
Hudson, A, and R Davies, 'Revealing the Hidden Mysteries of Holly', 20 December 2019, www.kew.org/read-and-watch/hidden-mysteries-of-holly
Tiryaki, D, et al., 'Psychrotolerant Bacteria Isolated from the Leaf Apoplast of Cold-adapted Wild Plants Improve the Cold Resistance of Bean (*Phaseolus vulgaris* L.) under Low Temperature', *Cryobiology* 86 (Feb 2019), 111–19
Wagner, J, et al., 'Cold Tolerance of the Male Gametophyte during Germination and Tube Growth Depends on the Flowering Time', *Plants* 6/1 (2017), 2

Chapter 6

Gols, R, and J A Harvey, 'Plant-mediated Effects in the Brassicaceae on the Performance and Behaviour of Parasitoids', *Phytochemistry Reviews* 8 (2009), 187–206
Kofidis, G, et al., 'Contemporary Seasonal and Altitudinal Variations of Leaf Structural Features in Oregano (*Origanum vulgare* L.)', *Annals of Botany* 92/5 (Nov 2003), 635–45
McAllister, H A, and R Marshall, *Hedera: The Complete Guide* (RHS, 2017)
Salisbury, A, et al., 'Careful Plant Choice Can Deliver More Biodiverse Vertical Greening (Green Façades)', *Urban Forestry & Urban Greening* 89 (2023), 128118
Soulis, K X, et al., 'Runoff Reduction from Extensive Green Roofs Having Different Substrate Depth and Plant Cover', *Ecological Engineering* 102 (May 2017), 80–9
Thomsit-Ireland, F, et al., 'The Impact of Green Facades and Vegetative Cover on the Temperature and Relative Humidity within Model Buildings', *Building and Environment* 181 (2020), 107009
Venn, S E, et al., 'Do Facilitative Interactions with Neighboring Plants Assist the Growth of Seedlings at High Altitudes in Alpine Australia?', *Arctic, Antarctic, and Alpine Research* 41/3 (2009), 381–7

Chapter 7

Berger, J, et al., 'The Appearance of Indoor Plants and their Effect on People's Perceptions of Indoor Air Quality and Subjective Well-being', *Building and Environment* 219 (2022), 109151
Berger, J, et al., 'The Impact of Plants on the Humidity of Naturally Ventilated Office Indoor Environments', *Journal of Building Engineering* 86 (2024), 108814
Dalling, J W, et al., 'Zombie Leaves: Novel Repurposing of Senescent Fronds in the Tree Fern *Cyathea rojasiana* in a Tropical Montane Forest', *Ecology* 105/3 (Mar 2024), e4248
Hetherington, A M, and F I Woodward, 'The Role of Stomata in Sensing and Driving Environmental Change', *Nature* 424 (2003), 901–8
Raux, P S, et al., 'Design of a Unidirectional Water Valve in *Tillandsia*', *Nature Communications* 11 (2020), 396
Xavier, V, et al., 'Leaf and Wood Functional Traits Explain the

Strategies Developed by *Byrsonima sericea* (Malpighiaceae) to Survive in Atlantic Forest Ecosystems under Water and Light Variations', *Flora* 308 (Nov 2023), 152386

Chapter 8

Clark, T, et al., '*Enkianthus* in Cultivation', *Meristems*, June 2011, www.yumpu.com/en/document/read/12231389/enkianthus-the-polly-hill-arboretum

Durand, M, et al., 'Sunfleck Properties from Time Series of Fluctuating Light', *Agricultural & Forest Meteorology* 308–9 (Oct 2021), 108554

Feng, Y-L, et al., 'Thermal Dissipation, Leaf Rolling and Inactivation of PSII Reaction Centres in *Amomum villosum*', *Journal of Tropical Ecology* 18/6 (Nov 2002), 865–76

Henry, H A L, and L W Aarssen, 'On the Relationship between Shade Tolerance and Shade Avoidance Strategies in Woodland Plants', *Oikos* 80/3 (Dec 1997), 575–82

Huang, D, et al., 'Morphological Plasticity, Photosynthesis and Chlorophyll Fluorescence of *Athyrium pachyphlebium* at Different Shade Levels', *Photosynthetica* 49 (2011), 611–18

Lv, Y, et al., 'Photochemistry and Proteomics of Ginger (*Zingiber officinale* Roscoe) under Drought and Shading', *Plant Physiology and Biochemistry* 151 (Jun 2020), 188–96

Murchie, E H, and P Horton, 'Acclimation of Photosynthesis to Irradiance and Spectral Quality in British Plant Species', *Plant, Cell & Environment* 20/4 (Apr 1997), 438–48

Nasrulhaq-Boyce, A, and M A Haji Mohamed, 'Photosynthetic and Respiratory Characteristics of Malayan Sun and Shade Ferns', *New Phytologist* 105/1 (Jan 1987), 81–8

Vidal, D, et al., 'Anatomical and Physiological Acclimation of *Fatsia japonica* Leaves to Irradiance', *American Journal of Botany* 77/9 (Sep 1990), 1149–58

Chapter 9

Burghardt, M, et al., 'Ecophysiological Adaptations of Water Relations of *Teucrium chamaedrys* L. to the Hot and Dry Climate of Xeric Limestone Sites in Franconia (Southern Germany)', *Flora* 203/1 (Jan 2008), 3–13

Cole, W S, et al., 'First Recorded Observations of Pollination and Oviposition Behavior in *Tegeticula antithetica* (Lepidoptera: Prodoxidae) Suggest a Functional Basis for Coevolution with Joshua Tree (*Yucca*) Hosts', *Annals of the Entomological Society of America* 110/4 (Jul 2017), 390–7

De Waal, C, et al., 'The Natural History of Pollination and Mating in Bird-pollinated *Babiana* (Iridaceae)', *Annals of Botany* 109/3 (Feb 2012), 667–9

Godsoe, W, et al., 'Divergence in an Obligate Mutualism Is Not Explained by Divergent Climatic Factors', *New Phytologist* 183/3 (Aug 2009), 589–99

Hacker, J, et al., 'Inflorescences of Alpine Cushion Plants Freeze Autonomously and May Survive Subzero Temperatures by Supercooling', *Plant Science* 180/1 (Jan 2011), 149–56

Jolly, A R, et al., 'Leaf Venation and Morphology Help Explain Physiological Variation in *Yucca brevifolia* and *Hesperoyucca whipplei* across Microhabitats in the Mojave Desert, CA', *Frontiers in Plant Science* 11 (2020)

Larcher, W, et al., 'Survival Types of High Mountain Plants under Extreme Temperatures', *Flora* 205/1 (2010), 3–18

Marks, R A, et al., 'Unexplored Dimensions of Variability in Vegetative Desiccation Tolerance', *American Journal of Botany* 108/2 (Feb 2021), 346–58

North, G B, et al., 'Contractile Roots in Succulent Monocots', *Plant, Cell & Environment* 31/8 (Aug 2008), 1179–89

Popova, O V, and D Golldack, 'In the Halotolerant *Lobularia maritima* (Brassicaceae) Salt Adaptation Correlates with Activation of the Vacuolar H^+-ATPase and the Vacuolar Na^+/H^+ Antiporter', *Journal of Plant Physiology* 164/10 (Oct 2007), 1278–88

Robberecht, R, and O Junttila, 'The Freezing Response of an Arctic Cushion Plant, *Saxifraga caespitosa* L.', *Annals of Botany* 70/2 (Aug 1992), 129–35

Wagner, J, et al., 'Winter Frosts Reduce Flower Bud Survival in High-mountain Plants', *Plants* 10/8 (Jul 2021), 1507

Warren, S D, et al., 'Directional Floral Orientation in Joshua Trees (*Yucca brevifolia*)', *Western North American Naturalist* 76/3 (Nov 2016), 374–8

Wierzbicka, M, et al., 'Micro-evolutionary Processes in *Armeria maritima* at Metalliferous Sites', *International Journal of Molecular Sciences* 24/5 (Feb 2023), 4650

Zlatić, N M, et al., 'Secondary Metabolites and Metal Content Dynamics in *Teucrium montanum* L. and *Teucrium chamaedrys* L. from Habitats with Serpentine and Calcareous Substrate', *Environmental Monitoring & Assessment* 189/3 (Mar 2017), 110

Chapter 10

Bepary, R H, et al., 'Biochemical Composition, Bioactivity, Processing, and Food Applications of Winged Bean (*Psophocarpus tetragonolobus*): A Review', *Legume Science* 5/3 (Sep 2023), e187

De Vahl, E, 'Historical Cultivars of *Allium cepa* L. (Aggregatum-Group) Introduced to Sweden 1830–1860', *Genetic Resources & Crop Evolution* 70 (2023), 2479–90

Jamalluddin, N, et al., 'Physiological Screening for Drought Tolerance Traits in Vegetable Amaranth (*Amaranthus tricolor*) Germplasm', *Agriculture* 11/10 (2021), 994

Lone, R A, et al., 'Adaptation of Winged Bean (*Psophocarpus tetragonolobus* (L.) DC.) to Drought Stress Is Mediated by Root-tuber Heat-shock Proteins and Specific Metabolites', *Current Plant Biology* 32 (Dec 2022), 100266

Olmo-Vega, A, et al., 'Physiological Responses of Three Pomegranate Cultivars under Flooded Conditions', *Scientia Horticulturae* 224 (Oct 2017), 171–9

'Rapunzel', www.grimmstories.com/en/grimm_fairy-tales/rapunzel (accessed July 2024)

'Spiked Rampion', Species Recovery Trust, www.speciesrecoverytrust.org.uk/spiked-rampion (accessed Jul 2024)

Yazdanpanah, P, et al., 'Seasonal Metabolic Investigation in Pomegranate (*Punica granatum* L.) Highlights the Role of Amino Acids in Genotype- and Organ-Specific Adaptive Responses to Freezing Stress', *Frontiers in Plant Science* 12 (2021), 699139

Index

Page numbers in **bold** refer to the main entries.

Acknowledgements

I am very grateful to Ana Moguš of Gush Art Studio for the beautiful images that adorn these pages, to Rosanna Fairhead for her careful copy-editing, and to Mark Fletcher and Kevin Hobbs for the original idea and ongoing support and guidance. Very special thanks to the garden design experts from around the world who took the time to discuss with me in detail the elements of garden design for different environmental conditions, including Mandy Barber (UK; www.incrediblevegetables.co.uk), Dr Tijana Blanusa (UK; www.rhs.org.uk/science/meet-the-team/environmental-horticulture-team/tijana-blanusa), Denis Dujardin (Belgium; www.denisdujardin.com), Sean Hogan (US; www.cistus.com), and Connor Smith (Netherlands; www.uu.nl/en/utrecht-university-botanic-gardens).

Source material

The illustrator used historic botanical illustrations as inspiration for the works in this book, as follows: *Addisonia* 107; Édouard André 201, 211; David Blair 126; *Botanical Magazine* 107, 186; *The Botanist's Repository* 158, 166; Maddalena Bouchard 203; Cistus Design 95; *Curtis's Botanical Magazine* 9, 123, 145; Sarah Anne Drake 27, 99, 105; Elisabeth Duisdeiker 73; Denis Dujardin 75, 135; *Favourite Flowers of Garden and Greenhouse* 128; *Flora von Deutschland* 16, 31, 120, 130, 179, 185, 199, 208; *Flora Graeca* 86; *Flowering Plants and Ferns of Great Britain* 20, 33; *Flore médicale des Antilles* 151; William Jackson Hooker 25, 143; *Incredible Vegetables* 215; William Kilburn 10, 103; *Köhler's Medicinal Plants* 17, 24; Jacopo Ligozzi 29; Carl Axel Magnus Lindman 49, 180, 183, 191, 205; Thomas Meehan 70; Maria Moninckx 210; Georg Oeder 131; Anne Pratt 7; Pierre-Joseph Redouté 18, 26, 122; G. Reid 79, 93; Jan Christiaan Sepp 85; Dorothea Elizabeth Smith 196, 213; Matilda Smith 65; John Stephenson and James Morss Churchill 58, 69; Pierre Jean François Turpin 43; Elizabeth Twining 164; Louis Van Houtte, *Flore des Serres et des Jardins de l'Europe* 76, 88, 110; Ferdinand Bernhard Vietz 200, 207; E. Weddell 67.

First published in Great Britain in 2025 by

Greenfinch
An imprint of Quercus Editions Limited
Carmelite House
50 Victoria Embankment
London EC4Y 0DZ

An Hachette UK company

The authorized representative in the EEA is Hachette Ireland,
8 Castlecourt Centre, Dublin 15, D15 XTP3, Ireland (email info@hbgi.ie)

Published in association with the Royal Horticultural Society.

A CIP catalogue record for this book is available from the British Library.

Hardback ISBN 978-1-52943-812-3
Ebook ISBN 978-1-52943-813-0

10 9 8 7 6 5 4 3 2 1

Text: Dr Amanda Rasmussen
Illustrations: Ana Moguš, Gush Art Studio
Design: John Round Design
Copy-editor: Rosanna Fairhead
Proofreader: Patricia Burgess
Indexer: Pauline Hubner

RHS Publisher: Helen Griffin
RHS Consultant Editors: Tijana Blanusa and Simon Maughan
RHS Head of Editorial: Tom Howard

The Royal Horticultural Society is the UK's leading gardening charity dedicated to advancing horticulture and promoting good gardening. Its charitable work includes providing expert advice and information in print, online and at its five major gardens and annual shows, training gardeners of every age, creating hands-on opportunities for children to grow plants, and sharing research into plants, wildlife, wellbeing and environmental issues affecting gardeners. For more information, visit: www.rhs.org.uk or call 020 3176 5800.

Printed and bound in China

FSC
www.fsc.org
MIX
Paper | Supporting responsible forestry
FSC® C008047

Papers used by Greenfinch are from well-managed forests and other responsible sources.